# HIGH STEEL

# HIGH STEEL

## Building the Bridges Across San Francisco Bay

Text by Richard Dillon
Photographs edited by Don DeNevi
and Thomas Moulin

Celestial Arts

Celestial Arts
P.O. Box 7123
Berkeley, CA 94707

First printing, September 1979

Printed in the United States

Cover design by Catherine Jacobes
Text design by Abigail Johnston

**Library of Congress Cataloging in Publication Data**

Dillon, Richard H.
    High steel

    1.   San Francisco—Bridges—Golden Gate Bridge.
2.   San Francisco Oakland Bay Bridge.      I. Moulin,
Tom, 1929-     joint author.      II   DeNevi, Donald P.,
joint author.      III    Title.
TG25.S225D54          624.5'5'0979461          78-72833

        4    5   6   7   —   01   00   99   98

Special thanks to Stuart Nixon
and his staff, The Redwood Empire Association
*Richard Dillon*


For Jeff Zafiris
not only because of his deep appreciation for
the Bay's two great bridges but also for his
conversations, advice, and friendship over the years.
*Donald P. DeNevi*

# CONTENTS

**H**ow many people have looked at the San Francisco-Oakland Bay Bridge or the Golden Gate Bridge and marveled at their beauty? The visual impact of these bridges is almost overwhelming at first sight. To the everyday commuter, however, they may be just a part of the highway system which takes him to work. But while serving as vital links in the commerce of the area and as flowing arteries for the commuters, these bridges are truly magnificent engineering feats.

To many they may be inanimate structures, but to others they are much more—they are living, dynamic and graceful. Perhaps this book can give the reader a new viewpoint and feeling for the uniqueness, wonder and achievement of these great tributes to the art and technology of bridge building.

Author Richard H. Dillon has captured both the challenge and excitement of planning and building the San Francisco-Oakland Bay Bridge and the Golden Gate Bridge. He writes of the difficulties in planning the projects experienced by Charles H. Purcell and Joseph B. Strauss, respectively, the driving individuals promoting the bridges. He continues into the exciting construction phase which required the cooperative effort of teams of engineers and describes some of the construction problems and accidents encountered in the erection of these giant structures.

But written words can only tell a portion of the story. It is the magnificent construction photographs by Moulin Studios that capture the excitement of building the bridges. They show in vivid detail the inexorable growth of these structures from the nearly invisible underwater substructures to the visually arresting superstructures.

As you view these superbly composed photographs, imagine the role of the bridge engineers behind the scenes, directing a caisson as it sinks in the bay, or a bucket as it places concrete, or a derrick as it lifts the high steel, or a spinning wheel as it lays wire forming the cables. Capture the founding of a caisson as it reaches bedrock, or the completion of a concrete pier, or the placing of the last high steel that completes the structure, or the laying of the last steel wire. This is the drama of bridge construction!

Bridge building is man's battle against the forces of nature—gravity, wind, current, and the expansions and con-tractions of temperature change. Indeed, the bridge must survive against the most enduring force of all—gravity—to serve its users. Just as the erection of towers and trusses is the bridge builder's struggle against gravity, the construction of the huge, deep caissons that form the foundations of the bridges is their struggle against current and water pressure. Foundations have never been constructed to such depths or to such large size or to carry such tremendous weights.

Let your eyes sweep over the photographs; see the whole structure, then look carefully at the details. Note the temporary devices used to erect the permanent structure and the discarded materials and equipment strewn across the deck. Scan the backgrounds: the ferries plying the Bay—they are gone now; the city skyline—there is now a forest of buildings; the barren hills—they are now developed. This is technology vividly illustrated by the construction of two of the most monumental bridges ever conceived and completed by man.

Then look at the completed structures and try to visualize each as alive and dynamic with beauty and grace. Follow the force of gravity as it flows downward to earth—and Mother Earth must ultimately carry all the forces of nature that develop on the structures. Feel the tremendous tension in the suspension cables as it transfers that force to earth through concrete anchorages at each end of the bridges. These are magnificent structures, built by man to serve man and in the process enhance the Bay area.

These bridges are a tribute to the engineers who planned, analyzed, and designed them, but also a tribute to the daring men who daily risked their lives in the struggle of building the bridges against the elements. These men were doing an exciting job and took danger in stride. Taking risks was their contribution to the building of the bridges and unfortunately a few contributed their very lives.

Today, the visitor or the commuter should view these wonderful structures not only as monuments to themselves but as monuments to those men whose courage enabled them to plan and erect these bridges to serve society ever after.

Charles Seim
*Chief, Operations Support, Toll Bridges*
*State of California Department of Transportation*

INTRODUCTION

# 1

# CONCEIVED BY A LUNATIC. . . . BUILT BY A GENIUS

Telegraph Hill                               Fort Point     Golden Gate              Pt. Bonita

View of the Entrance to San Francisco Bay from Yerba Buena Id.

In an official bulletin, the U.S. Census Bureau pronounced the American frontier to be finally dead in 1890. It was a case of premature burial. Most free land was gone by 1890, the continent was crossed and re-crossed by railroads, and the conditions of a settled society now prevailed from Atlantic to Pacific.

Historian Frederick Jackson Turner rightly predicted that the main push and energy spent earlier by Americans in conquering the continent would move into new "channels of agitation" elsewhere, but he did not give enough consideration to a force which was running almost parallel to the westward movement—the Industrial Revolution, a nationwide concern for science and technology.

The most obvious manifestation of the new "frontier" came in the spatial thrust into aviation with the Wright brothers, Wiley Post, Charles Lindbergh and others. Almost simultaneously it arose in the massive program of public works which, during Franklin D. Roosevelt's New Deal, developed into a major campaign of FDR's war against the Great Depression.

The keynote to the philosophy of this new frontier can be found in Roosevelt's remark that "We can no longer escape into virgin territory; we must master our environment." Looking back to that time, you could say that the phrase had an ominous ring. But in the depressed '30s, before the term ecology was coined, these brave words were a call to action. The people responded by applauding Roosevelt's dream and it became not only good business, but actually patriotic, to build—and to build *big*. Here, at last, was another frontier, a brand new kind of pioneering. Great dams were constructed for flood control, water power, land reclamation, irrigation and recreation. With one huge exception, the TVA (Tennessee Valley Authority), it was a western phenomenon. Even its modest beginnings, in 1911, were at Roosevelt Dam in Arizona.

During the 1930s the movement peaked with Bonneville Dam in Oregon; Fort Peck Dam in Montana; Grand Coulee Dam in Washington; and, particularly, Boulder Dam (or Hoover Dam) on the Colorado River between Nevada and Arizona. Boulder Dam became symbolic of them all, but the technological frontier did not culminate in forebays and penstocks and reservoirs, but, rather, in the chords and trusses and towers and cables of the two greatest bridges in

the world. Incredibly, they were both built at the same time, in the depths of a depression, by the one city which itself symbolized hope and optimism during those dark days, San Francisco.

The San Francisco-Oakland Bay Bridge and the Golden Gate Bridge were true phenomena. They tested—boldly challenged—the perceived limits of engineering at the same time they belatedly extended the old western transportation frontier, which had reached an earlier climax in 1869 with the driving of the last spike of the transcontinental railroad at Promontory, Utah.

San Francisco's bridges solved a centuries-old transportation and communication problem. They contributed to a postwar boom by making possible a ring of suburban settlement around the peninsular city. They made the City readily accessible to both commuters and visitors, and facilitated a whopping 39 percent Bay Area population increase in the years 1940-1947 alone.

The two-part San Francisco-Oakland Bay Bridge is a mixed breed in its engineering design. The west Bay crossing is a handsome, if conventional, suspension span. The east Bay crossing's cantilever span and line of through-trusses is impressive, to be sure, but downright homely.

The Golden Gate Bridge, the Cinderella of spans, is a masterpiece. Its soaring grace actually enhances the beauty of its natural setting, something which man-made objects rarely do. It has been a revelation to even its most severe critics, winning over all of them by the time of its completion. The largest work of art in history, dwarfing such an example of combined art-and-engineering as the Eiffel Tower, it is the greatest sculpture ever dreamed.

Important for their material benefits, the bridges transcend more utility in their spiritual value. They were inspirational in nature, their construction heartening Americans during the darkest year of the depression, 1933, when enough laborers to populate a small nation (12 million workers) were unemployed. After the spans were completed in 1936 and 1937, the inspiring work was carried on by the city in a party atmosphere typical of San Francisco, with the delightful world's fair of 1939-1940—the Golden Gate International Exposition—on a man-made atoll in San Francisco Bay, Treasure Island.

Only the attack on Pearl Harbor and the shock of the

GOLDEN GATE INTERNATIONAL EXPOSITION

on San Francisco Bay

1939

(Courtesy Herbert Wasserman)

United States' entry into World War II could, temporarily, dash the buoyancy of San Francisco, as it hauled itself out of gloom with such giant public works. (Even in wartime, the bridges—at least the Golden Gate span—remained symbolic of the west and, indeed, of the whole country. Briefly, the Golden Gate Bridge was even dubbed an "Arch of Victory" after the V-J Day triumph over Japan in 1945.)

Like the ancient Egyptians, San Francisco's bridge builders were not content with mechanical triumphs of engineering and architecture. Their common goal was art *and* beauty. In the case of the Golden Gate Bridge, which easily eclipses its sister span (where the latter is impressive in its magnitude, the Gate span is breathtaking in its dignity and gracefulness), they sought to create one of a new Seven Wonders of the World, replacing the loss in antiquity of the Colossus of Rhodes. The guardian of the Golden Gate is perfect in scale, flawless in line.

The bridges have been likened to a single "symphony in steel" by the San Francisco historian Rev. John B. McGloin, S.J. Sausalito's best-selling author of the late 1920s and early '30s, Frederick O'Brien, hailed the future "highway in the sky" to Marin County as "an incredible road of shining metal that will hang from cliff to cliff." The Golden Gate Bridge has inspired poetry as well as prose, fiction as well as non-fiction. Poems were written by Robin Lampson and John May, in *Scribner's* and *The American Scholar* respectively, to honor the building of the grandly sculptured monument spanning the narrows, which John C. Fremont called "Chrysopylae," from a fancied resemblance to Constantinople's Golden Horn. The pathfinder's pompous Greek was promptly—and properly—anglicized to "The Golden Gate."

Poet Cristel Hastings won the hearts of Marinites by likening the Gate span to a magic portal to "summer's wide domain," and expressing the hope that its towers would be as enduring as the redwoods whose lofty height they imitated. Chief engineer Joseph B. Strauss wrote a poem, *The Mighty Task Is Done*, to his bridge and found that at least one of his laborers, ironworker Howard McClain, was also a poet who took the span as subject matter. Later, a children's book would place a troll beneath its deck, making it a troll bridge as well as a toll bridge. Alistair McLean would write a paperback thriller about it.

"Golden Gate Bridge Under Construction," an oil which now hangs in the White House, was painted by Ray Strong, a student of Maynard Dixon, the fine California muralist and unofficial painter-in-residence of Boulder Dam. The bridge inspired powerful graphics by the woodcut artist, Mallette Dean, and by Otis Oldfield. However, it fell to photographers to really capture San Francisco's great bridges.

Peter Stackpole took dramatic shots, as did Ted Huggins for Standard Oil Company, while George Dixon buzzed the spans periodically to provide aerial shots for Tidewater Associated Oil Company. But the men who best recorded the bridge construction, and converted photo-documentation into art, were commercial photographers Gabriel, Irving, and Raymond Moulin of San Francisco. They were the official photographers: Moulin, *pere et fils*, were granted contracts by the State of California to document the day-by-day progress of construction. Typically, Gabriel and his sons refused to accept the job as routine. Sensing the challenge and drama of men-and-steel, brawn-and-iron, they did the best work of their careers.

**G**abriel Moulin was born in San Jose in 1872 of French and German parents. He moved to San Francisco at age 12 and at 14 dropped out of Lincoln Grammar School to become a $12-a-week assistant to the City's leading photographer, I. W. Taber, spending his evenings in the lab, learning the secrets of developing glass plates and making prints.

By 1894 and the Midwinter Fair in Golden Gate Park, young Gabriel was an accomplished photographer. In 1898 the Bohemian Club asked him to photograph its annual summer encampment in its Russian River grove of redwoods. The dark, shadowy sequoias were difficult subjects, but young Moulin tamed them by his mastery of darkroom techniques—and the exploding of enough flash powder, it is said, to illuminate Mammoth Cave. Some critics say his best pictures were accidental, when the smoke from earlier flashes drifted through the well-lit trees, to make pictures which were marvelous mood studies. In any case, he deserved the title "The Redwoods Photographer." In his obituary, the *Examiner* would state, "He made the redwoods famous." His distinguished pictures of the trees brought countless visitors to the "redwood empire" and made it almost as famous as the city which he had adopted as his own.

Gabriel Moulin was well ahead of his time. Perhaps he anticipated the "f64 Group" of Bay Area photographers (Ansel Adams, Edward Weston, Imogen Cunningham, Alva Lavenson, and others) which revolutionized photography in the 1930s. He closed down his lens opening (hence the f64) and lengthened his exposures to ten seconds (to minutes, in some cases) in order to create landscapes whose sharp definition was astounding. He and his sons used this general technique of small opening/long exposure during the building of the bridges, but here the technique occasionally betrayed them. Surf sometimes turned to cotton wool (like the white water of the Merced River in the excruciatingly slow shutter speeds of Yosemite's early photographers), and the *Malolo*, unless dead in the water under the Golden Gate Bridge, would be a blur as she dashed for the open sea. And as for the China Clippers which soared past the bridge towers like clockwork—almost impossible.

Just before the turn of the century, Gabriel joined the firm of R. J. Waters & Co., but quit on April 18, 1906, the day of the San Francisco earthquake, when Waters refused to rescue the negatives in their fire-threatened Ellis Street studio. Gabriel began his independent career by shooting dramatic scenes of the City lying in ruins. He was soon established as San Francisco's most creative cameraman. Small wonder that he was appointed photographer of the Palace of Fine Arts during the Panama-Pacific International Exposition in 1915, where he pioneered the use of Lumier color plates.

Gabriel Moulin was a creative photographer in the best sense. In his 60-year-long career, he sought perfect composition and sharp images. He turned away from the stylish soft focus, the self-consciously arty and posed work, which was imitative of the vogue in "real" art. He experimented with many photographic techniques, testing various exposures and regulating contrasts in his pictures, but he did little cropping and refused to embellish or re-touch weak negatives or use other gimmicks of the trade. Instead, he resolutely sought perfection in line and dimension, light and shadow, balance, composition and framing. He searched for the ideal vanishing

point and depth of field. Most important, he instilled this discipline into his photographer-sons, Irving and Raymond.

Irving had joined his father's business in 1923, but Raymond worked as a photographer for *Pictorial California* and Californians, Incorporated, a travel firm (even sailing to the Orient and South Pacific as a cruise ship photographer) before joining Moulin Studios about 1930. The firm was already internationally famous for commercial work, the "pictorial recording" of the development of California industry, as well as for landscapes and portraits. While Irving won San Francisco retail advertisers over from traditional artwork, line drawings, etc., to photography, Raymond designed an aerial camera to crystallize the City's essence.

Gabriel was 61 when bridge construction began. He took many panoramic views but left the difficult—and dangerous—work aloft to his sons, especially Raymond, a fearless photographer. With as little vertigo as a mountain goat, he scrambled over towers and cables, or teetered on girders suspended in thin air, while lugging camera and equipment. Most of the pictures in this book are Raymond's; in a sense the volume is a tribute to him.

Gabriel died in 1945. Irving retired in 1957. In 1974 Raymond turned the studio over to his son, Tom.

The period which architectural critic Allan Temko calls "the heroic age of American bridge building" reached its apogee during 1933-37, with the construction of the two San Francisco bridges. It is this pair of great spans, not the City's Manhattanized skyline nor the spilled-spaghetti freeways of Los Angeles, which have become a metaphor for the pioneering spirit of the American West, noted in 1890 by Frederick Jackson Turner, but now in modern dress.

Luckily for San Francisco, both bridges were conceived and built by engineers who refused to be intimidated by the sheer technical challenges that faced them. They saw beyond the difficulties of span and stress and load, sensing the need to harmonize with the natural environment of one of the most beautiful settings in the world, San Francisco Bay. Temko properly described the bridges as being not only mighty structures but "courageously beautiful (and) democratic works of art."

Gabriel Moulin

Raymond Moulin

Irving Moulin

Emperor Norton
(California Historical Society)

**S**ince the line between madness and genius is the proverbial razor's edge, it is tempting to say that San Francisco's bridges were conceived by a lunatic and built by a genius. The genius was a visionary in a hard hat (the "five-foot giant" of columnist Herb Caen), named Joseph Baermann Strauss. The lunatic whose plan was anticipated by a whole platoon of Californians was the mad, self-styled Emperor Norton.

William Walker, editor of the San Francisco *Herald* and most extravagant proponent of Manifest Destiny (via his filibustering expeditions into Baja California and Nicaragua), may have been the first person to propose a great Bay bridge. It was 1851, scarcely seven years after the first rude plank bridge had been thrown across an unnamed creek in the village still being called Yerba Buena. Observing the success of the 2,000-foot Clay Street Wharf, Walker suggested a causeway be built to the Oakland shore with a pontoon bridge on anchored barges in the deep water area. Mayor John Geary is believed to have supported the idea.

In 1853, Dr. D.G. Robinson built a model of a fanciful Bay bridge as part of a dramatic extravaganza. But this was *ante bellum* science fiction. Historian George R. Stewart calls the whole bridge idea a fantasy, though "masquerading as genuine," since it was far beyond the technical or financial resources of the area at that time.

On February 13, 1856, 13 years before the transcontinental railroad reached the coast, State Senator W. H. McCoun introduced a bill to grant a railway and wagon road right-of-way across the Bay. On March 9 the San Francisco *Daily Alta California* editorialized against it as a plan to enrich speculating monopolists, citing it as one of the "inventions of keen political adventurers always infesting legislatures and ready-charged with schemes for their own advancement." Although eleven senators voted for it, the Senate postponed action on the bill indefinitely, and it simply vanished without a trace.

However, Stephen Massett, alias James Pipes of Pipesville, a vaudeville entertainer, kept the idea itself alive in 1856, when he set to music a poem by Charles Mackay, *Clear the Way!* He retitled it *The Song of the Oakland Bridge.* The most popular wit of the day, Lt. George H. Derby (alias Squibob, alias John Phoenix), promptly penned a parody which also burlesqued Coleridge's *Ancient Mariner.* Derby's doggerel referred to "hydraulic characters," pile drivers, and "men of suction," and exhorted San Francisco to "pump the Bay dry!" The bridge idea died as a tired joke.

Joshua Norton was a greedy speculator who lost his mind as well as his fortune while trying to corner the rice market in a city thronged with Chinese. Later he was adopted by the fun-loving City as a kind of scruffy municipal mascot in his new role of Norton I. He issued many imperial decrees, but his most famous was one dated August 18, 1869, which revived the bridge idea. He issued a command for a San Francisco railroad bridge to be built not only to Oakland but to Sausalito as well. (And, indeed, to a dead end on the Farallon Islands, 30 miles out to sea!) The Oakland *Daily News* dutifully published his decree:

> We, Norton I, *Dei Gratia,* Emperor of the United States and Protector of Mexico, do order and direct . . . that a suspension bridge be constructed from the improvements lately ordered by our royal decree at Oakland Point to Yerba Buena (Island), from thence to the mountain range of Saucilleto (sic), and from thence to the Farallones . . . Whereof fail not under pain of death!

Norton, in part, may have been crazy as a fox, or perhaps the tool of San Franciscans already leery of Central Pacific Railroad Company (CPRR) power. There was a real fear that the CPRR would build a "Chinese wall" along the Contra Costa (Alameda) shore to shut City interests off from the east, except on the terms of the railroad barons. Norton's cockeyed plan was followed by several San Francisco papers backing a plan to locate the transcontinental railway terminal in the City, not on Central Pacific's East Bay shore. George F. Allardt headed a commission of engineers which estimated the cost of a causeway plus two drawspans, either of "preserved" (creosoted) wood or of iron and stone, at $17 million. It would run from Mission Bay to the Oakland shore, carrying two railroad tracks, a horse-car track, a carriageway and a walkway for foot traffic. The causeway was chosen over a suspension span for reasons of safety and utility as well as economy.

Promoters hoped to place hotels and shops in the bridge's towers a la London Bridge or the Ponte Vecchio.

These plans were the precursors of such schemes as a tourists' elevator for sightseeing in the original plans for the Golden Gate Bridge, and Joseph Bazzeghin's 1937 dream of roller coaster rides on both San Francisco bridges.

Although the public scoffed at the idea, the San Francisco *Daily Evening Bulletin* and the *Morning Call* joined the *Daily Alta California* (now perhaps a "prisoner" of the Central Pacific) in advocating that the bridge be built, at municipal expense. An anonymous correspondent to the Sacramento *Daily Union,* "Pioneer," wrote that public furor for a bridge was all "bosh," that it had no popular support. It was all a CPRR scheme to make a cats-paw of San Francisco. "Pioneer" was sure that the cost figure, now $5 million, was much too low, and predicted great difficulty for pile drivers penetrating perhaps 25 feet of mud on the Bay bottom. Also, he wrote, the earthworks would be swept away by gales and tides. The writer suggested that the railroad go ahead with its planned crossing from Ravenswood, near Palo Alto, to Niles, separated by only two miles, not seven miles, of water.

Without endorsing the measure, San Francisco Supervisor Charles R. Story on October 23, 1871, submitted to the Board a resolution to hold a special election to vote on the issuance of construction bonds. Central Pacific and other railroads would have free passage on the span, from Mission Bay's Point San Quentin, for 12 years. This last item was enough for Supervisor James Adams. He moved to postpone indefinitely action on the measure. Supervisor Alexander Badlam then moved to refer it to committee. Jokingly, Supervisor Stewart Menzies recommended its referral to the Hospital Committee, since it was so sickly an idea. The chairman facetiously accepted, and it died there for lack of action.

Some bridge advocates wanted a first-class suspension span, carrying both a railroad and a military highway from Telegraph Hill to a "goods terminal" on Yerba Buena Island, connected with the mainland by a causeway. This would not only let first-class (large, ocean-going) sailing vessels pass under it, but would be an honor to San Francisco, and unlike an ugly causeway, would "elicit praise from all the world." But they wanted Central Pacific to pay for it, not the City. Governor Leland Stanford, of the Central Pacific/Southern Pacific "Big Four" and practically the inventor of the term "conflict of interest," in the west, now was all for the bridge—

*if* the City would first purchase the bonds. Oakland opposed the project, fearing it would be bypassed as a rail terminus.

An anonymous pamphlet, *The Railroad System of California* (1871), pushed the Bay bridge idea and interest ran high. The Marysville *Daily Appeal,* for example, ran three stories that fall and winter, and the Sacramento *Daily Union* six stories. Sacramento, or at least the *Union,* opposed the project, ostensibly because the causeway might render the harbor inaccessible for first-class ships. The paper reminded its readers that professors Benjamin Pierce and George Davidson of the U.S. Coast Survey denounced the plan because it would cause the Bay to shoal and ruin the harbor's future. It also reminded them that the House had had to defeat a bill which would have turned Goat Island (Yerba Buena) over to the Central Pacific as its terminus. The real reason for the *Union's* opposition was seen in its battle cry, "Bay and harbor of San Francisco, the pride of all California, not for sale to the railroad company!" (No wonder the CPRR banned the sale of the *Union* on its trains.)

Many San Franciscans agreed with the Sacramento newspaper. They perceived the project as a veiled theft of more public property and treasure by railroad magnates. They saw to it that the idea died a quick death. It did not revive until 1906, except for one brief moment in 1891, when one Henry Langrehr came up with a Bay crossing idea.

The Central Pacific, metamorphosed into the Southern Pacific, finally drove a low-level bridge, a trestle, across the shallowing, narrow, south Bay in 1906. It ran between Dumbarton Point and the Palo Alto sector of the San Francisco Peninsula. Not until 1927 did the low auto span, the Dumbarton Bridge, parallel it. That same year saw the first high-level bridge, the precursor, in a sense, of both great San Francisco spans. This was the Carquinez Bridge, the first to be built to withstand earthquakes. In 1929 the San Mateo Bridge was completed—a low-level structure with a drawspan like the Dumbarton.

But the deep waters, powerful tides and swirling current, lashing winds and blinding fogs—not to mention the post-1906 fear of earthquakes—so put the fear of God into engineers that serious proposals for such seismically "vulnerable" bridges, as skeptics put it, were thwarted until World War I.

Joseph B. Strauss
(California Historical Society)

In 1868 there was newspaper talk of a company being formed to link San Francisco with Marin County by a bridge to Lime Point. But not till World War I did people in Marin, like James Wilkins and Thomas Allen Box, and San Franciscans like Richard J. Welch, institute action. Supervisor Welch introduced a resolution to authorize a survey (1918) of the geological conditions of the site.

San Francisco's brilliant, Irish-born City Engineer, Michael M. O'Shaughnessy, asked engineer Joseph B. Strauss, half-jokingly if he would be interested in bridging the Golden Gate. He warned, "Everybody says it can't be done. And that it would cost over $100 million if it could be done." Strauss replied, "I think it can." He estimated a cost of $25-35 million. And he was not joking.

From a perch on Fort Point bluff, Strauss studied the breech in the Coast Range. Although no center anchorage was possible in that mile of whitecaps covering a 300-foot depth of water, he vowed to span the Gate. He asked the Board of Supervisors for copies of Welch's charts, soundings, and topographical data. The City, to be on the safe side, invited plans from two other engineers of Strauss' caliber. One failed to respond; the other quoted a cost of $77 million. Strauss' (1921) figure was $27 million.

Joseph Strauss was the only serious candidate for the job. It was no wild flight of imagination to him. In an anonymous pamphlet which he apparently wrote with O'Shaughnessy, titled *Bridging the Gate,* he echoed the sentiments of Frederick Jackson Turner: "In such a product of the Great Golden West, America could build a Peace Memorial that would fitly commemorate the close of the World War . . . Aside from its commercial value and financial attractiveness, and its great practical value, it will represent a crowning achievement of American endeavor and will constitute the greatest structure in point of magnitude and span ever erected."

Feisty and combative like many men of small stature, Strauss (barely five feet tall) thought big. His graduate thesis in civil engineering at the University of Cincinnati had been a bridge across the Bering Sea. He was an inventor (and a poet) as well as an engineering genius with several hundred bridges all over the world. He was as confident as the fellow engineer he once quoted: "I'll build a bridge to Hell if they'll give me enough money to do it." He was a titan in terms of energy, ideas, and vision. Already, he had revolutionized rail-road bridging with his concrete-weighted bascule designs. He would more than make history with the Golden Gate Bridge; he would immortalize himself.

Ironically, the opposition which quickly formed against his project proved, ultimately, to be fortunate for the Bay Area. It gave Strauss time (with the prodding, perhaps, of his engineering assistant, Clifford E. Paine) to radically change his original design, an ugly, hybrid cantilever-suspension bridge, squatting heavily in the Gate and cluttered with the worst elements ever borrowed by either architect or engineer from the Eiffel Tower ("the junkman's Notre Dame"). He eventually substituted the clean lines of the classic suspension span which was actually built.

Opposition came not only from San Francisco esthetes but also from environmentalists in Marin (like this writer's father, Captain William T. Dillon, Secretary of the Sausalito Chamber of Commerce), who were distressed by Strauss' case for speculative "development" in the beautiful county via the bridge. The chief engineer preached "tremendous growth of this wonderful district" and seemed to suggest housing tracts on the edge of (and perhaps inside of) Muir Woods. Strauss prophesied all kinds of prosperity—but chiefly increased property values—"in the wake of the building of this great example of Western supremacy."

The War Department was expected to oppose Strauss' project out of a fear that, in time of war, a bombed bridge would bottle up the harbor. Although hostile attorneys twisted his words in a hearing before the Army Engineers, changing his description of flexible steel cables and towers into "a rubber bridge" (for laughs), Strauss won the Secretary of War's permission to build in 1924.

Meanwhile, O'Shaughnessy, more than enthusiastic in 1919, had cooled off. Strauss later said the Irishman simply gave up, telling him, "Your bridge is all right. But you can never convince people it can be built." Indeed, Strauss found that his bridge was an unwanted orphan. When he was introduced to the Commonwealth Club it was as "the father, mother and wet nurse" of the unbuilt bridge. But he won support when he spoke effectively of the solid rock on which it would be built; of the fact that it was *not* on an earthquake fault; that bridges resisted quakes better than any other structures; and that because of its long center span, his bridge would be the most flexible on earth. He also predicted the bridge would pay for itself in tolls. And when a doubtful O'Shaughnessy later asked him, "How long will your bridge last?" Strauss shot back—"Forever!"

Although Marin County, the Redwood Empire Association, the California State Automobile Association and other organizations backed the project by 1930, Strauss found himself engaged in a "Ten Year War." Taxpayers' suits took the constitutionality of the district and its projected bond issue (to finance the structure) all the way to the Supreme Court even after the district was incorporated in 1928, in another victory for Strauss.

The chief engineer chose as consultants O.H. Ammann and Leon S. Moisseiff. The district's directors added the Dean of the College of Engineering at the University of California, Berkeley, Charles Derleth. Paine remained Strauss' chief assistant. In 1929 Strauss opened a downtown office in San Francisco, a field office at Fort Point, and a satellite of the latter across the channel. By 1933, he had a 100-man engineering force at work on plans and the inspection of structural materials in shops on both coasts.

As his geologist, Strauss chose Professor Andrew C. Lawson of the University of California, the "discoverer" of the San Andreas earthquake fault. He thereby provoked another conflict. The press called it "The Battle of the Geologists."

Pitted against Lawson was Stanford's Dr. Bailey ("Earthquake") Willis, famous for being able to predict quakes with some degree of success. The white-bearded, 76-year-old professor picked up a joking remark by a south tower deep- sea diver about the rock bottom being "as soft as plum pudding." (According to Time, the same diver reported mermaids lounging in underwater caves.) Willis began to call the greenish serpentine rock of the Gate "pudding stone". He said it was so unstable that the whole structure would be endangered.

Lawson's retort was "Buncombe!"—professorial talk for just plain "Bunk!". He accused his rival of needlessly scaring people. Strauss later admitted, "It was disconcerting to be caught halfway between professors." Bureaucrats of the Public Works Administration (PWA) lost their nerve over the "pudding stone" and Strauss got not one red cent of federal money. But the canny chief engineer won back the timid by drilling 100-foot test holes every 25 feet over a sea bottom area the size of a ten-acre farm, and then ordering an exca-

vation of 35 feet into bedrock—100 feet below the waves—for the footing of his south pier. He could not quite convert Willis into a supporter but at least got him to confine his criticism to a grumpy, "Time will tell."

The soil tests and boring samples proved to be as satisfactory as Strauss' new design and specifications. He chose the simple suspension span design more out of economy than esthetics, for the bridge would have to be built entirely in the private sector. And he certainly was not after any world records for bridge size. He recalled, "I did not intend to attempt to construct the longest span in the world. On the contrary, I tried to make the span length as short as possible." The Secretary of War's final permit specifications of 1930, for a 4,200-foot span with a vertical clearance of 220 feet at mid-span and 210 feet at the towers, actually determined for Strauss the great height of his towers, 746 feet above mean sea level.

Opposition picked up as the bond election of November 4, 1930, approached. A citizens' committee fought a heated campaign (and was joined by 13 prominent engineers) to urge defeat of the project because the structure was believed to be both physically and financially impossible. A banker called the project "an economic crime." The mighty Southern Pacific, parent of the ferryboats doomed by the bridge, waffled. (Against the bridge in 1930, it gradually reversed itself by 1932.) But Strauss was a fighter. He battled back with an information office and a speakers' bureau. Endorsed by labor unions and civic organizations, the bond issue of $35 million was passed by 145,057 votes to 46,954.

Although Strauss immediately arranged for surveys and conferences on the bridge approaches with Army and state highway engineers, and the Board asked for bids on June 17, 1931, another round of litigation prevented work during 1932. As the district's funds began to run out, a controversy over the legality of the bonds' rate threatened to cripple the project. Strauss moved promptly. He cut through the tangle of legalistic red tape by going directly to the Bank of America's A.P. Giannini. The bull-necked Italian hardly resembled Strauss physically, but was a man after his own heart. In spite of an unfavorable bond market, he pledged his bank's support in the purchase of $3 million of the bonds at the legal 5 percent rate. He commented, simply, "San Francisco needs that bridge. We'll take the bonds."

# 2

# THE SAN FRANCISCO-OAKLAND BAY BRIDGE

WEST CROSSING      YERBA BUENA      EAST CROSSING

During the late 19th and early 20th centuries, Yerba Buena Island's grassy slopes were often the destination for picknicking San Francisco day-trippers. The San Francisco-Oakland Bay Bridge's cantilever span would cross the deep channel between the island and Southern Pacific's busy Long Wharf and Oakland Mole Terminal.

Even before Strauss sized up the treacherous Golden Gate, agitation for a trans-Bay bridge was growing. In 1916 a War Department hearing on several proposals brought no action. But in 1928 another Army hearing coincided with the support of President Herbert Hoover. With Governor C.C. Young of California, the President appointed the Hoover-Young Commission to make preliminary studies. That same year the State legislature created the California Toll Bridge Authority and authorized construction of a highway across San Francisco Bay.

The Bay Bridge did not have the long uphill fight of Joseph Strauss' "baby." It was an official project of the State, built by the Department of Public Works for the Toll Bridge Authority and to be operated by the Division of Highways. "The Feds" stood behind it, also, like beaming grandparents. With Hoover's blessing, it easily won not only Army and Navy approval but funding by the new Reconstruction Finance Corporation (RFC). In 1932, the RFC agreed to purchase up to $61,400,000 worth of revenue bonds.

Still, even though it would run from a big city to a pretty-big city, unlike the Gate span which tied San Francisco only to the dairy ranches of Marin, there were doubting Thomases. For one thing, it would be 8.4 miles long, 12.5 miles including approaches. And 4.5 miles of that was either an overwater span, per se, or through Yerba Buena Island. It would have to be three times longer than the current world's longest, the Firth of Forth Bridge. The opposition also predicted that the job would require the greatest expenditure in history on a single structure. They were right. The price tag came to a whopping $78 million. Worse, the foundations would have to penetrate a great depth of water and mud in hopes of finding safe beds while, on either side of the structure, ran the San Andreas and Hayward earthquake faults.

Just as in the case of the Golden Gate Bridge there was worry among San Franciscans that the bridge would mar the beauty of the Bay. And, once again, they were right. The original design was as bad as Strauss' first span, a huge pair of cantilever bridges joined by Yerba Buena Island. The mass of girders, in the words of Bay historian Harold Gilliam, would have been about as esthetic as Chicago's El—but 20 times as big. Fortunately, citizens refused to be bulldozed, and the plans evolved into a handsome suspension-span west

Bay crossing, with the homely cantilever and truss section of the east Bay crossing partially screened from peering eyes in City highrises by the mass of Yerba Buena Island.

Strauss' opposite number was Charles H. Purcell. He was the State's Chief Highway Engineer and also Secretary of the Hoover-Young Commission. Like Paine, he was quiet and did not make the papers as did Strauss, but he was very competent and experienced. Born in Nebraska in 1883, he had been forced to drop out of Stanford for lack of funds but graduated from the University of Nebraska in 1906. He won for himself a fine reputation as a civil engineer, building roads and bridges, until Governor Young chose him as State Highway Engineer.

Assisted by Charles E. Andrew and Glenn B. Woodruff, Purcell in 1930-31 made preliminary designs and studied borings. His consulting engineers were Moisseiff and Derleth of Strauss' staff, and his geologist was Dr. Lawson. Another consultant was Ralph Modjeski, son of the famed Polish patriot and actress, Madame Helena Modjeska. (Strauss had been Modjeski's principal assistant in 1899-1902.)

With plans and specifications completed in late 1932, invitations to bid were sent out and contracts awarded by April 1933. Two of the seven major contracts comprised the largest structural steel orders ever placed with a single firm, the American Bridge Company.

The first construction work was begun in May of 1933, though the formal groundbreaking by Hoover and Governor Frank Merriam had to wait till July 9. The Governor ceremoniously hefted a clod or two in his gilded shovel and the ex-President orated about "the greatest bridge ever made by human hands." President Franklin D. Roosevelt pressed a telegraph key at 12:47 which detonated charges on Rincon Hill, on Army Point of Yerba Buena Island, and at the foot of Oakland's 14th Street. Hardly had the echoes of the explosions died away before excavating was begun and, by November, there were gaping gashes in the island, awaiting piers and anchorage.

Purcell knew the limits of his art. No one could lay a single suspension bridge across the 9,250 feet of deep water between Rincon Hill and the island. He would have to build not one bridge but three, all rolled into one. It would be easy to connect Yerba Buena Island with the Oakland shore by a conventional bridge. But the west crossing would have to be two 2,310-foot suspension bridges (with 1,160-foot sidespans) connected at a common anchorage. (Only one such tandem bridge existed, a small one built in Prague in 1841 across the Moldau River made famous by the composer Smetana.) To form a mid-stream superpier, he would have to plant a brand-new, man-made, island in the Bay. Thus, the suspension span would have five piers but only four towers. Pier W-4, to be built by Moran & Proctor Construction Company, would be a double anchorage.

Luckily for Purcell, he found (like Strauss) an underwater ridge of rock. It ran all the way to the island, but lay beneath 100 feet of water and another 100 feet of mud. The bridge's foundations would have to be deeper than any ever built. And he could not dump rock to create his island; such loose fill would never provide his cables with the grip they needed on Mother Earth. Nor would a concrete atoll serve. His anchor pier would have to be a high island of concrete to take the live and dead loads of the span and survive possible earthquake shocks. It must support a deck 261 feet above busy shipping lanes. At 92 feet by 197 feet, the size of the caisson he intended to use to build it, and extending 210 feet below water, it would not only be higher than the City's tallest skyscraper, the 33-story Russ Building, it would be the largest pier in the world.

Since the "island" was the key to the success of the entire bridge, Purcell sought the help of 70-year-old Daniel E. Moran, *the* expert on deep-water foundations. They pored over contour maps, sections (profiles) of the Bay bottom, and boring cores. The 200-foot combined depth was *twice* the range of "sand hogs" (laborers) working under pressure, but they finally came up with a plan.

Workmen laid the keel, as it were, for "Moran's Island" when they built a pneumatic or compressed air caisson on the shipways of the Moore Dry Dock Company in Oakland. The Moran and Proctor caisson was half the size of a city block. Its four timber walls (ending at the bottom in sharp cutting edges of steel) sheltered, topside, five rows of eleven vertical steel cylinders, 15 feet in diameter. The cells were open at their bottoms but sealed at their tops with hemispheric, airtight domes, which gave the caisson its buoyancy. Towed to the site, the great caisson was wedged between two working

platforms. By pouring concrete around the cylinders, the structure was sunk slowly in place as workmen extended upward its walls and cylinder tops to keep them always above the water level of the Bay.

When the caisson was supported by the Bay floor, the welded domes were removed in sequence, and clamshell buckets dropped down the tubes to dredge up the mud trapped between the watertight walls of the great box, 6,800 pounds of muck at a gulp. After the caisson rammed the cutting edges down into solid rock, divers cleaned the rock's surface. Next, sharp-pointed gads were dropped down the chutes to break up the rock to permit leveling for a good footing. The square chamber at the caisson's bottom was then filled with concrete to make the caisson an integral part of the pier. The watertight concrete mat was extended up into most of the cylinders for only 35 feet. Beyond, they were left filled only with water. Corner cylinders were plugged completely.

Pier W-4, which grew out of the caisson, consumed more concrete—165,000 cubic yards—than the Empire State Building or six Russ Buildings. "Moran's Island" is the size of a 40-story building covering a city block, far bigger than the largest of the Pyramids of Gizeh.

Tour boat operators today will tell you that "Moran's Island" is the tomb of from 4 to 7 workmen who fell into the fresh concrete and were buried alive. According to the unofficial historian of the Bay Bridge, Charles Seim, the actual total was—zero. The story is a dramatic one, but it is totally untrue.

The other piers of the suspension span were formed in the same fashion as W-4. At 240 feet, W-3's maximum depth was even greater than that of the anchorage-island, but no serious difficulties were encountered as new records were set for subaqueous construction.

Purcell and Moran must share honors with Bill Reed, chief diver, for the bridge's success. He was paid $15,000 a year, a fabulous Depression salary, plus a dollar-a-foot for each dive he made. He earned every penny. The west Bay piers ranged from 100 to 240 feet in depth, the 23 east Bay piers from a mere 50 to 242 feet at one end of the cantilever span. Six of the piers penetrated deeper under water than any on the globe.

As early as 1930, when he worked on the Southern Pacific railroad bridge across Carquinez Strait, Reed was

From the clock-towered Ferry Building at the foot of Market Street, a great fleet of San Francisco ferryboats fanned out across the Bay. By the 1930s, the system carrying 50 million passengers a year was taxed to its limits, and the old dream of a bridge spanning the Bay was not only revived but, this time, brought to reality.

At ground-breaking ceremonies for the Bay Bridge on July 9, 1933, almost all eyes were on the speaker, but one young celebrant was more interested in the bridge's official photographer.

A bucket brigade of Bay Area ground-breakers momentarily "watched the birdie" as they planted a ceremonial redwood tree.

famous. (He even made the sports pages in December of 1929 when he listened to the Cal-Stanford Big Game—while working underwater.) The *Chronicle* described him as "the intrepid San Francisco diver" and the "noted submarine adept." Reed's advertisements for his business at 50-60 Steuart Street, San Francisco, were hardly models of modesty: "Hazards! are many and oft-times fatal in submarine diving engineering . . . We delight in hazard undertaking because we have the equipment, experience and men to do them. No matter how difficult, consult us first on your submarine diving, marine salvaging, wrecking, underwater construction and examination problems."

Reed had taken the Carquinez job after other divers refused to dive to depths of 149 feet to inspect concrete pier footings. In 21 years, the veteran claimed dives to 206 feet. He had raised wrecked ships and broken-down bridges after beginning his career in Toronto hauling sacks of cement on a waterway project as a boy. When the regular diver got sick, Reed volunteered to go down, though he had never suited up before. He loved it and began to work all over the United States. He was the first diver to reach the sunken submarine S-51 in 1925 and the first to bring up a body. In 1927 he came to San Francisco and salvaged a cargo of copper for Luckenbach Steamship Company which other divers said was impossible. He designed his own equipment, including a diving bell for undersea photography and, from 1927 to 1930, salvaged shipwrecks with the help of only two men. He also made the initial soundings for the Golden Gate Bridge.

According to the *Chronicle* reporter who interviewed him for the November 1, 1930, paper, Reed was reluctant to talk about the dangers of his work. But he admitted to some close calls, such as the times he had had to cut his own air hose to extricate himself from sunken ships.

No diver, not even Reed, could work steadily at the Bay depths. With pressure mounting to 100 pounds per square inch (from the 14.7 pounds of normal atmospheric pressure), he could only dive for 10 to 15 minutes at a time. But that was enough time for him to inspect the critical underpinnings of Purcell's bridge. Reed's hands and fingers became Purcell's underwater eyes. There was no light on the bottom; he had to work blind. Reed substituted his marvelously developed sense of touch for sight.

A delicate job for Reed was pier E-5, where ten feet of bottom was excavated beneath the caisson. It was his job to crawl under 56,000,000 pounds of concrete to see if dredge buckets and water jets had done a good cleaning job. In 21 days, Reed made 23 deepwater dives. According to Purcell, they varied from 170 to 185 feet. In just one day, he made three 170-foot dives, a new record for the Bridge.

Typical of Reed's hazardous duties was his straightening of a canted caisson hung up on an obstruction in the mud, and tilting dangerously. Work was suspended while Reed's fingertips groped along the "hull" of the caisson, found a projecting rock, planted a dynamite charge—and got out of there. When the detonation removed the offending bump, the caisson swung neatly back into place.

Each time he dove, Reed took his health, perhaps his life, in his hands. The age-old danger of the bends required that he be rushed into a decompression chamber after each dive, where the air pressure would gradually be decreased to normal. There was always the chance of an accident as he squeezed between thousands of tons of steel or concrete in tide-swept, choppy, waters. He also found the Bay bottom to be a junkyard of anchors, cables, and jetsam. He had to pick his way blindly through the debris and, once, had to cut his lifeline when it became badly tangled in a rusty old cable. He was hauled to the surface on his air hose.

Across the marshy shallows of the Oakland shore, with its pickleweed and duck blinds, Purcell ran a series of simple plate girder spans and through-truss spans from a viaduct and *terra firma*. The dry land piers of Yerba Buena Island were easy enough, as were the 20 shallow water piers, the latter founded on timber piles and built within sheet-pile cofferdams. By February 1935, the two-boom traveler derrick advancing along the deck had replaced falsework with permanent bents (supports) and truss panels as far as the deep water immediately behind the island.

Purcell chose to cross the channel with a huge cantilever span (1,400 feet long, 22,500 tons, and 191 feet above the water), the largest and heaviest in the U.S., but cheaper than a suspension bridge in that location. It was built by a guy-derrick traveler—a gigantic metal spider—creeping forward and devouring steel from barges below. Half of the span was up by January 1936. The main problem was pier E-3 at the

The Bay Bridge was a project close to the heart of former President Herbert Hoover and he was the natural choice for keynote speaker at the ground-breaking ceremonies: "Now let work begin. This marks the physical beginning of the greatest bridge ever erected by the human race."

Governor Frank Merriam of California shoveled some soil to formally begin construction of the bridge. San Francisco's boutonniered mayor, James "Sunny Jim" Rolph, beamed his approval over the Governor's shoulder.

San Franciscans came down the slopes of Rincon Hill to see ground broken—and history made—on the site of a future pier and anchorage behind the Haslett Warehouse.

The keys to successful construction of the Bay Bridge were its deepwater piers. They rose from floating caissons which were sunk into place. Moore Dry Dock Company's role in the bridge's construction was essential—it built and launched the great caissons from its yards and ways.

deep end of the cantilever structure. It was 300 feet down to bedrock. It would be difficult to sink a caisson there, impossible to send a diver—even Reed—that deep. So Purcell planted the pier on a solid substratum of hard clay and sand. Even so, at 242 feet, E-3 was the deepest pier in the world. A 16-story building set alongside its base would not break the water with its lightning rod.

To support the west Bay's suspension span cables, four steel towers were erected on piers W-2, W-3, W-5 and W-6. ("Moran's Island," W-4, was an anchorage, of course, without a tower.) W-2 and W-6 rose 474 feet, but the other two soared to 519 feet in order to bring the bridge floor to its highest point, 216 feet above harbor traffic, at the anchorage pier. Each tower was built of two cellular columns, or legs, of silicon steel joined by X-shaped crossbraces and horizontal supports, or struts. The flexible towers could take a load of 65,000,000 tons. Their tops could bend four to five feet out of line in obedience to load and temperature extremes. Sections of the towers, averaging 50 feet in height and 75-80 tons, were shipped by train from the east and towed by tugs to their piers on railroad car floats.

The usual creeper traveler, a derrick which clung to the surface of a tower and moved upward like a second-story man, could not be used because of the design of the tower legs. Instead, the engineers of the American Bridge Company used the center cell in each leg as a well or shaft to hold a 180-foot vertical steel mast. On top of it was a small rotating mast fitted with the crossarm of a hammerhead derrick. It was served by a stiff-leg derrick set between the bases of the two columns. After four tower sections were placed, one atop the other, the hammerhead was raised 50 feet and the operation repeated.

The hammerheads "committed suicide" by erecting guy derricks which promptly dismantled them. These guy derricks then placed the grooved cable saddles on the tops of the columns. The saddles were the largest single-piece castings ever used in bridge work. When the first saddles arrived, it was March 1934, and a bloody waterfront strike was paralyzing the Embarcadero. Purcell preferred to take risks rather than submit to costly delays, so he ordered the ship to "dock" at a bridge pier where a stiff-leg offloaded its critical cargo and passed the saddles up to the guy derrick above. (The guy

derricks were left in place to help with cable spinning.) Towers were built one at a time, each taking three 5-day weeks of two 6-hour shifts of workmen. Not until each entire 5,000,000-ton tower was erected were temporary bolts replaced with 110,000 hot field rivets.

At the ends of the sidespans, the cables would rest on short inclined steel columns which rotate on huge pins at the bottom. These were the cable bents which provided the support for the cable between the end towers and the anchorage. Beyond them the backstay cables extended to the anchorages, great concrete monoliths sunk deep into the rock and entombing tiers of steel eyebar chains which grasped the splayed-out cable strands firmly. Since the cables stretched and shrank with load and temperature variations, the bents rocked on great pins at their bases.

On the anchorage pier, W-4, each cable, splayed into strands like an unraveled rope, was tied by eyebars to a 40-foot high steel A-frame. Its legs were anchored to "Moran's Island" by chains of eyebars extended to girders imbedded 135 feet deep in the concrete, to handle the tremendous pull of the cables. Collars called splay castings prevented "unraveling" of the cables beyond their proper splay points.

Each of the four 28¾-inch cables consisted of 17,464 galvanized (.195 inch) wires in 37 strands of 472 wires each. In all, 18,500 tons of spliced wire (thousands of miles of it), with a tensile strength of 220,000 pounds per square inch, was used on the Bay Bridge.

Before the "aerial spinning" could begin, two 10-foot catwalks for workmen had to be slung from the towers in curves corresponding to those of the future cables. To cut down on weight, wind resistance and fire hazard, American Bridge engineers used strips of galvanized chain link fencing stapled to timber crosspieces and "paved" with fine mesh hardware cloth. Sections were hoisted to the tower tops, spliced together, and slid down the catwalks' support ropes or cables. Storm cables held the footwalks safe from excessive wind sway and the crosswalks connecting the parallel catwalks acted as stiffeners. Naturally, the catwalks were tempting to would-be "bridge monkeys" who had had a few drinks. One of the first to make his way to Yerba Buena Island and back on the cyclone fencing was George Christensen of San Francisco.

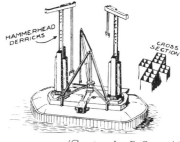

(Courtesy Ian R. Samuels)

The tugs, barges, pile drivers and derrick barges operated in close teamwork with the gigantic pneumatic caissons to create the deep-water piers of the Bay Bridge. But the real drama took place on the bottom of the Bay, where divers like Bill Reed and Lloyd Evans risked their lives daily in submarine inspections of the piers.

Frugal Purcell later cut the catwalk support ropes into short lengths for suspenders for the deck's trusses, and used the storm cables as permanent hand ropes atop the finished cables for inspection walkways.

Timber gallows frames (squared-off, inverted "U"s) on the catwalks carried sheaves for the powered, endless, hauling ropes on which the spinning wheels rode. Electric lights on the footwalks allowed spinning to be done at night, weather permitting, though (curiously) it was not carried on during Saturday afternoons or Sundays.

To begin spinning a cable, men first made fast the ends of wire on two reels to the anchorage. The wire was then looped over a large pulley-like wheel with two grooves. Hauled across the Bay, the wheel laid down four wires, each trip, to be looped around horseshoe shaped castings called strand shoes, which were connected by pins to the anchoring eyebars. The two lower, inert, wires were the anchored "dead wires," the upper ones the "live wires" advancing from the

reels continuously along the growing cable. Each group of four strands was called a "set-up." As each reel was emptied, the ends of the wires were spliced to the ends of the next reel, making endless loops of wire.

An innovation by Purcell was to add companion wheels working from the opposite end of the span. Both were fed by reel stands of four active spools, or reels, at each anchorage. The wire passed from them to the spinning wheels via counterweight floating sheaves which maintained proper tension on it.

The main cables were completed and pneumatic squeezers were busily compacting them (by January 1936) and changing their shapes from hexagonal to circular. Six screwjacks, mounted on a yoke or collar surrounding each cable, advanced and squeezed with great pressure, then ran a few turns of temporary seizing wire around the cable to hold its shape for the wrapping machines. The latter were also yokes, but with bobbins instead of jacks, which crawled

The first major change in the Bay's appearance since the construction of the Key System's curving electric railroad trestle, ferry slips, and train sheds was the straight line of piers across the shallows of the *contra costa* to the deep water between the Key System pier and Yerba Buena Island. (In the distance, the Golden Gate Bridge's north tower is up and Joseph Strauss' much-battered trestle juts out into the Gate.)

along the cables they surrounded and wound a tight spiral of soft, annealed galvanized wire around them for permanent protection. Near the 596 cable bands, from which would hang the 1,192 suspender ropes holding the deck trusses, the wrapping was done, painstakingly, by hand, with men tucking the wire under the grooved edges of the cable bands and caulking the remaining space with lead wool. Lastly, the wrapping was sealed with paint.

The westernmost of the twin suspension cables was spun first, then the equipment moved to the east span on November 12, 1935. When its cable was completed on January 20, 1936, an average of 128 tons of wire had been laid per day, easily beating the George Washington Bridge's old record of 99 tons.

The stiffening trusses supporting the bridge's two decks, and preventing twisting of the span from live load (traffic) changes, ranged from 75 to 200 tons each. They were barged to the bridge in one of the most difficult peacetime

"fleet operations" in our history. Tidal currents were too powerful for the engines of tugs to hold the barges in place long enough to attach the lifting apparatus. (They could not anchor for fear of dragging—and breaking—power and communication cables on the bottom.) Finally, four widely spaced anchors were planted where they would not foul cables. Lines attached to them were run to the drums of hoisting machines on a special anchor barge. This could be planted securely in place and the loaded barges moored to it. The hoisting lines, four sets of falls attached to struts laid across the cables, required but ten minutes' work by the four hoisting engines at the nearest tower base to raise and position the trusses. But men had to guide them into perfect alignment and bolt together their chords (horizontal members). This factor, plus the vagaries of wind and wave below, kept progress down to two units per day, three on a very good day.

The heavy truss units were placed in position in a strict sequence, with deliberate gaps left in the span in order to avoid an unbalanced pull and, thus, excessive tower deflection and cable distortion. (Stringers, curbs, concrete, etc., were later added in a staggered order, also for the same reason.) Daily measurements showed a maximum tower deflection of only three inches, thanks to Purcell's precautions. All of the 146 two-panel units were raised between December 18, 1935, and April 30, 1936.

Once the trusses were all in place, three 14-ton guy derricks began to run on the upper deck level, placing the balance of the crossbraces and the floor steel. They were followed by riveters who replaced the bolts in the chord joints.

A routine operation, which was critical nevertheless, and even "ceremonial," temporarily caused a problem for the engineers. The time came for the twin halves of the span cantilevered over the channel (east of Yerba Buena Island) to meet and be bolted together, completing the final link of the bridge. Meet they did, as the closing members were hoisted into place by the huge insect-like guy derrick travelers. But they refused to fit! As workmen scrambled over the steel like bugs, the engineers found that the pinholes of the lower chords were short of matching by 10 inches those of the upper chords by 13 inches.

Apparently, a cool sea breeze was shrinking one side of the span on a sunny warm day. To connect the lower chords, the entire structure was jacked 10 inches westward from the expansion joint of pier E-4. When the chord holes matched, a triumphant gang of hard hats swung a battering ram to drive home the last chord pin. The 3-inch gap in the upper chords was closed by jacking and by extension of these members. And perhaps the weather changed (it always does on San Francisco Bay) to help, rather than hinder.

**W**est Bay and east Bay crossings were connected by the world's largest and longest vehicular tunnel, a double-decker drilled and blasted for 540 feet through Yerba Buena Island. Three "pioneer" tunnels were first bored, then broken out by excavation to create an arch which was concreted. Then the core was excavated and the spoil dumped into the Bay to partly create Treasure Island. So big

was the finished tunnel, 52 feet, 8 inches high and with a clear width of 65½ feet, that a 4-story house could have been constructed inside of it.

**P**aving of the two bridge decks was done in short, alternating strips of concrete poured into wooden forms. The upper level became a six-lane "automobile boulevard" for cars and buses. The lower deck was occupied by three (north side) lanes for trucks and by twin standard-gauge tracks for the electric trains of the interurban Key System. (They were removed in 1958 and the two decks made into five-lane, one-way, suspended highways, the lower level eastbound, the upper deck westbound.)

The last rivet was driven on October 23, 1936. On November 12, after three years and seven months of hard but rapid (two months ahead of schedule) work by an army of men putting in 54,850,000 man-hours of work, completion was official. Several hundred thousand people participated in a three-day celebration. Some 14 naval ships belched smoke and man-made thunder in a cannonading salute as 250 Navy planes flew overhead in formation.

At the Oakland bridgehead, Rabbi A. A. Stern offered prayers for the 24 workmen who had lost their lives on the bridge. (Purcell did not think the number was excessive.) Ex-President Hoover spoke once again. Purcell, his staff, and a hard hat representing his thousands of buddies, took bows. Governor Merriam then used a dirty acetylene torch to cut a golden link in a silver chain and *Time* preserved his rather banal observation: "This bridge is not the product of a day." As the chain hit the pavement at 12:27, 1,500 pigeons were released into the blue sky. According to poetically licensed *Time*, some harbor seals popped their heads above the surface of the Bay to see what was going on and then (as if on cue) added their barks to the cheers of the throng. Overhead, the winds toyed with the smoky calligraphy of a sky-writer, his plane lettering "The Bridge Is Open."

Dignitaries raced across the bridge to repeat speeches on the San Francisco side. FDR pressed a golden key in the White House which triggered green lights, and six long col-

The deeply cut anchorage site on Yerba Buena Island afforded photographer Moulin a splendid view of progress on the west Bay crossing in November, 1933. An enormous caisson was being jockeyed into position to lay the foundation for pier W-6. Unbuilt as yet in the busy shipping lanes were W-5 and the great double-anchorage pier, W-4.

A ferryboat cuts between the caisson of pier W-6 in the foreground, and the pneumatic domes of mighty W-4, nicknamed "Moran's Island" for the underwater-foundation expert Daniel E. Moran, who made the world's largest pier possible.

Workers who dismantled the wooden forms from the concrete of pier W-6 near Yerba Buena Island revealed what appeared to be a two-turreted concrete monitor of some futuristic navy, riding at anchor.

Between the shallow-water piers north of the Key System pier and a concrete pylon being framed on Yerba Buena Island's Army Point, the east Bay crossing was supported by the deepest footing of the bridge. This allowed for the largest and longest (1,400 feet) cantilever span in the country.

umns of cars shifted into first with a nervous grinding of gears you could hear on Rincon Hill. The three eastbound and three westbound phalanxes of autos moved out. Pedestrians lined both rails, but only on that day, for there were no provisions for them, unlike the Golden Gate Bridge.

The largest and most expensive bridge in the world had a gargantuan appetite. It had swallowed 200,000 tons of structural, cable and reinforcing steel; 1,000,000 cubic yards of concrete; 1,300,000 barrels of cement, and 22,000,000

rivets (give or take a few). It cost 78 million depression dollars and 24 lives.

The decks of both bridges were engineered with a flexibility to accommodate traffic loads and temperature changes. Both have the capacity to move up and down over 20 feet; however, their normal movement is approximately 4-5 feet.

Was it worth it? Absolutely. It probably could not be built today at any price. There are not enough skilled workmen in the labor force and not enough inflated dollars in our coffers.

Tower W-2 soared above the masts and booms of freighters anchored off the waterfront and high above the tallest buildings of downtown San Francisco.

The glory of the San Francisco-Oakland Bay Bridge is the series of silver suspension span towers. Viewed from a derrick on the Embarcadero anchorage, a tower built on pier W-6 soared 474 feet as hammerhead derricks put finishing touches on its top. (June, 1934)

The Pacific Fleet parading past the huge concrete bulk of "Moran's Island," flanked by the towers of piers W-5 (left) and W-3.

28

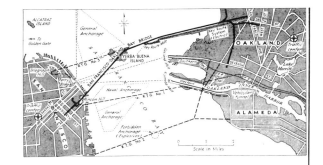

**I**t is said that a visitor, a distinguished gentleman, was loaded with statistics and information, all true, by Bay Bridge guides. They told him that it took 18 percent of all the steel fabricated in the United States in 1933; that it required enough timber to build houses for a town of 15,000 people; sufficient reinforced concrete to rebuild downtown San Francisco, or create 35 copies of the Russ Building or the Los Angeles city hall; that the 71,000 miles of wire locked into its cables would circle the globe more than three times and, if laid end to end would stretch three-quarters of the way to the moon!

The visitor accepted all these astounding facts serenely. But then he saw a 50-ton cantilever member being lifted into place. Next he watched as a 200-ton stiffening truss unit was snatched up off a barge and deposited like a jigsaw piece into the span 200 feet above his head by slim hoisting lines and skinny suspender ropes. He continued to listen patiently to his proud bridge guide, but shook his head in disbelief. "I cannot believe my eyes. I cannot believe you. It just cannot be so. It's too marvelous."

The Bay Bridge has an inferiority complex because of its homely east Bay crossing vis-a-vis the beauty and popularity of Strauss' fairy princess of a span. Visitors cannot flock to it on foot, though many drive across it just for the splendid view. It is not favored for suicides, having drawn only 126 so far compared to the Golden Gate Bridge's 680 or more. Still, it *was* blessed by the Pope (in 1936, when he was Cardinal Pacelli) and it *was* named one of the seven engineering wonders of the world by the American Society of Civil Engineers in 1956.

What more can a bridge ask? Except to be loved, like the Golden Gate Bridge.

Only a ferry bound for Sausalito broke the calm of a quiet day as the west Bay crossing towers neared completion.

The 33½-foot-high cable-bent on top of pier W-1 (left), inclining 5 degrees to the west, gripped and supported the great cables in saddles (resembling those of the tower tops) at the point where the sidespan ended and the cables' backstays extended 863 feet into the bedrock of the gravity anchorage.

In a sense, the lofty tower on pier W-6 (foreground) is a memorial to the first of 24 men who died during construction of the Bay Bridge. In December, 1933, diver Lloyd J. Evans, working at a depth of 112 feet removing pins from cables anchoring the caisson of pier W-6, suffered excruciating pains in his legs. He died after an eleven-hour fight to save his life from the dreaded bends.

Beneath overcast skies, a setting sun threw long shadows across the streets "South of Market" as the towers of the west Bay crossing neared completion.

Matson Lines' sleek *Malolo,* bound for Hawaii, sails by the historic ferries *Alameda* and *Oakland,* while the first ropes to support the cable catwalks of the Bay Bridge were being strung.

The first step in spinning the cables was the draping of wire ropes over the towers to support walkways. The curves corresponded to the sag of the future cables on which the men would be working.

Even with elevators, "high iron" work meant men scrambling on steel girders far above the ground, hence their nickname—"bridge monkeys."

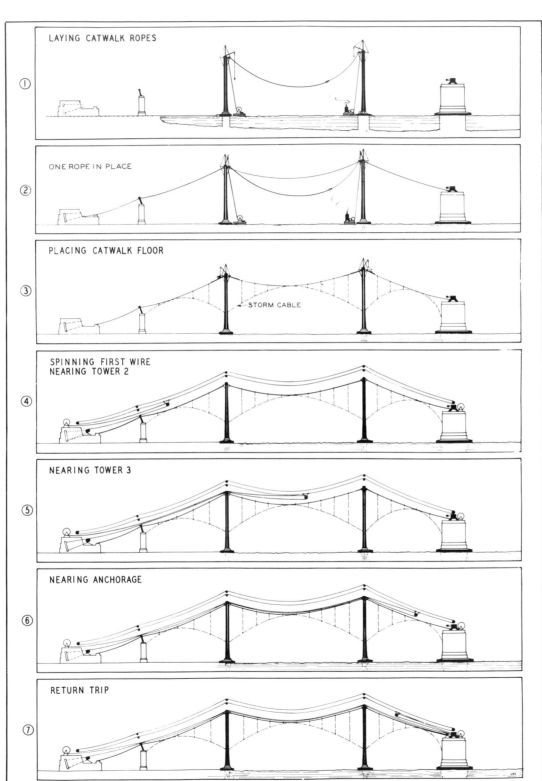

*Cable spinning.* On the Bay Bridge spinning wheels made a complete transit from anchorage to anchorage, while on the Golden Gate Bridge the spinning carriages went only to mid-span, where their bights of wire were manually shifted to opposing wheels and then returned to their home anchorage.

High above the Embarcadero's traffic, hardhatted workmen tinkered with a gallows frame through which a cable would be spun, while a co-worker gazed at them anxiously from the catwalk below.

The mast and booms of the freighter *Montanan* seemed to gesture toward the workers securing the "pavement" of the Bay Bridge catwalk.

Workers appeared to be kneeling in space as they unrolled hardware cloth to cover the cyclone fencing and finish the catwalk. Rope handrails for safety were still to come.

While their buddies finished the south catwalk, two hardhats took a break on a newly erected crossbridge. These crossbridges made movement from one catwalk to another rather easy and, in addition, lessened the sway caused by strong winds through the Golden Gate.

With the nonchalance of his breed, a "bridge monkey" walked backward up the south catwalk, dragging a roll of hardware cloth, while the photographer clung to a sky-high perch to make the shot.

39

High steel workers were daring men, but not foolish. When they posed for Moulin hundreds of feet in space, they kept to the center of the narrow catwalk because the posts and the wire safety ropes were not yet up.

From the tower on pier W-2, the loosely grouped strands sagged to the concrete of W-1 and on to the anchorage before they were compacted by the squeezing machines into circular cables.

The enormous pull of the bridge cables was distributed at the anchorages by splaying or fanning out the strands from a collar (right) and attaching each one to an individual strand shoe (left) pinned to parallel eyebars.

On Yerba Buena Island and at San Francisco's Rincon Point, ranks of eyebars deeply imbedded in the concrete anchorages stood ready to grip the strands of the mighty cables being spun on the west Bay crossing's twin suspension spans.

This rickety looking spinning wheel (one of a pair) threaded together the elements of the greatest bridge in the world.

Men and wheels, spinning webs of steel across the Bay, draped the cable wire over the saddles at the top of the towers.

The spools of strand used for the suspender ropes dwarfed the workmen who handled them.

The colossal grip of the jacks in a squeezing machine not only compacted the many strands into a unified cable, but insured a perfectly circular cross-section. (Note that *all* Bay Bridge workers did not wear hard hats for safety.)

Ferries and a solitary steam schooner plodded across the Bay together as the bridge cables finally began to take shape.

Even without finished cables, the west Bay crossing's twin suspension spans were a beautiful sight as their catwalks, gallows frames, storm cables and topmost guy derricks were lighted at nightfall.

Passengers aboard a Pan American *China Clipper* had a sweeping view of the bridge in progress: the suspension spans of the west Bay crossing, the east Bay cantilever-truss span, and the Yerba Buena tunnel linking the two. (Clyde Sunderland)

By New Year's Day, 1936, the cantilever span east of Yerba Buena Island was taking final shape at the same time the island's suspension span anchorage and the connecting tunnel were progressing.

The double-decked Yerba Buena Island tunnel, the largest vehicular tube in the world, was built to connect the west Bay crossing (the suspension span) with the east Bay crossing (a combined cantilever and through-truss bridge). The geometric plans and repeated archings of the portal for the upper roadway suggested the period's art deco vogue. Excavation by Fairbanks power shovels created the lower level, and the "spoil" was deposited in the Bay to form Treasure Island, another monument to art deco during the Golden Gate International Exposition of 1939-1940.

While cables were being spun on the west Bay crossing, the east Bay's cantilever-truss span on the other side of the Yerba Buena tunnel was being fitted together section by section.

Moulin's documentary photographs, capturing men and the geometry of inanimate steel in curious juxtaposition, often transcended their subject matter and became art.

To connect the two halves of the cantilever span, workmen swung a great battering ram to drive home a pin through the matching holes of the closing members.

The last gap in the entire bridge lay between the two sections of the cantilever span east of Yerba Buena Island. The guy derricks lifted into place the closing members— only to find they did not quite fit! The entire east half of the structure had to be moved westward 10 inches by powerful jacks at the expansion joint of pier E-4.

When the two portions of the cantilever span finally met, the Bay Bridge was considered to be "complete," although it was unfinished. Moulin captured the scene on a perfect sunny morning.

Shaded by the super-
structure the deck of the
east Bay crossing pro-
gressed across the
shallows to Oakland's
*terra firma.*

As the bridge neared completion, the graceful lines of the suspension crossing were extended by a curved approach from the island tunnel to the cantilever span between piers E-2 and E-3.

Homely, perhaps, when compared to its suspension span extension westward, or to the Golden Gate Bridge, but the east Bay crossing was nevertheless handsome after its completion.

As the Bay Bridge was being prepared to receive its roadway, work proceeded on the Golden Gate Bridge's north tower, beyond Coit Tower in the distance.

The truss units for the decks of the suspension span, weighing from 75 to 200 tons, were hoisted into place from anchored barges below.

Sections of stiffening trusses were hung from socketed cables called "suspender ropes," or simply "suspenders," in a strict sequence. The gaps were carefully planned to avoid unbalanced strains on the cables and tower tops.

At last, the stringers and crossbeams were in place, ready to receive the concrete forms of the roadway.

The uniform of the day was overalls and a choice of headgear—cloth cap or U.S. Navy "swabbie" hat—for the painters of the suspender ropes holding the deck's stiffening trusses.

On a gusty day, painters appeared to get more pigment on themselves than on the suspender ropes, as they dangled on bosun's chairs.

As time wore on, the powerful stiffening trusses created more and more complex geometric shapes.

The concrete of the bridge's two decks was reinforced by trusswork and "re-bars," (reinforcing steel bars). Note the line of white tiles at left, traffic lane markers.

Fresh concrete was poured in place from buggies in as carefully choreographed an operation as the placing of the deck trusses. The paving was done in a staggered sequence to prevent sudden weight imbalances and consequent strain on the towers and cables.

The symmetry and clean lines of the tandem suspension spans were most evident in this last Moulin portrait of the west Bay crossing without a "live load" (traffic).

A lone official pondered the future of the beautiful bridge, gleaming in its new coat of silvery paint.

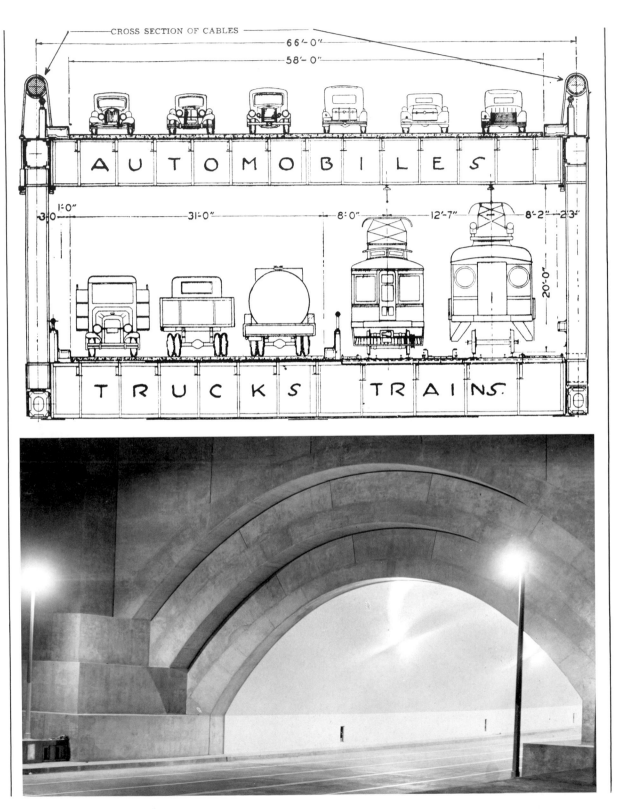

The original plan for the Bay Bridge had automobiles only on the upper deck, trucks, buses, and trains on the lower; later, the rails were removed and the upper deck made one-way, San Francisco bound, and the lower one-way, Oakland bound.

CROSS SECTION OF CABLES

66'-0"

58'-0"

AUTOMOBILES

1'-0"

3'-0"    31'-0"    8'-0"    12'-7"    8'-2"   2'-3"

20'-0"

TRUCKS    TRAINS

The completed tunnel, aglow with lights.

With the cooperation of the U.S. Navy, the Bay Bridge presented a fabulous light show the night before it opened to traffic.

On opening day, crowds gathered early at the San Francisco approach to salute the drivers who would make history on the new route across the Bay.

Wearing the soft caps and whipcord breeches of the '30s, the Bay Bridge's 19-man squad of California Highway Patrolmen lined up across the deck in front of the patrol cars and Harley-Davidsons just before the bridge opened to the public.

The Sacramento Northern Railroad, the Key System and Interurban Electric Railway (Southern Pacific lines) operated electric trains on the lower deck of the Bay Bridge. In 1958, a Key System train made the last run over the tracks of the Bay Bridge. Interurban Electric Railway's "red trains" test car (seen below) made a run before the start of regular rail service across the bridge.

The crowd at the Oakland toll plaza was a large one on the first day of traffic. It included the marching band from U.C. Berkeley, which waited patiently for the ceremonies to start.

Pedestrians participating in opening day ceremonies crowded the approaches and offramps.

From a plane high above the west Bay crossing on opening day, cars looked like insects. The City was lightly veiled in fog.

# 3

# THE
# GOLDEN GATE
# BRIDGE

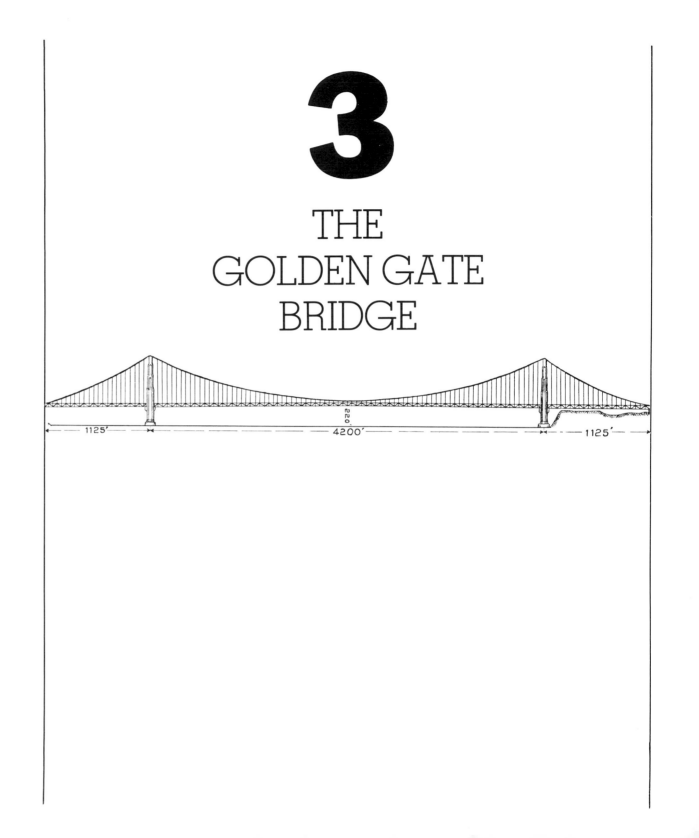

1125'    4200'    220'    1125'

Lawyers and litigants had consumed so much time that most Golden Gate Bridge low bidders were released from their contracts, and new bids were received on October 14, 1932. The total, $23,843,905, suggests how prices had declined in the depression.

Too impatient to wait for groundbreaking ceremonies, Strauss got preliminary construction under way on January 5, 1933, and excavation by the 12th. Actually, work had already begun (December 22, 1932) on an access road to the north pier site on Lime Point. Even earlier, of course, divers like Bill Reed had made Gate soundings from the drilling barge skippered by Captain Rudolph Petterson.

The February 26, 1933, groundbreaking at the Presidio's Crissy Field was attended by 200,000 people, ten times as many as would attend the Bay Bridge's similar ceremony. Governor James B. "Sunny Jim" Rolph and Mayor Angelo Rossi spoke. The only hitch in the proceedings was the grounding (by fear) of a flock of carrier pigeons intended to spread the good news. They huddled in their cages until small boys, armed with sticks, prodded them into reluctant flight.

At last the 15-year-old struggle to be allowed to build his bridge was over for Strauss. He later said, "It took two decades and 200,000,000 words to convince people that the bridge was feasible; then only four years and $35 million to put the concrete and steel together."

Strauss' two great Roebling cables of pencil-thin wires would handle a vertical load of 210,000,000 pounds. With a safety factor of 2.6, they would support a deck crowded with cars, its sidewalks jammed with people. He allowed for a 16-foot sag due to climatic and live load (traffic) extremes, figuring the temperature range to be from 30° to 110° F. The latter was farfetched for the site, but the thermometer sometimes dropped to 25°, and laborer Pete Williamson used to knock icicles from the cables. The clearance would be 210 feet above mean high water at the piers and 220 feet at mid-span—just enough, it would be seen, for the *Queen Elizabeth I* to fit on a high tide.

The keystone to the entire structure was the south pier, planned 1,125 feet out in the heavy swells and white-capped tides of the Golden Gate. It was essential to have quiet water in which to build the pier, via a caisson, so Strauss planned a

great "fender," a closed oval wall of concrete, inside of which his men would work in a metaphorical millpond. The pond would not only be shaped like a football, it would be the size of a football field: His "underwater stadium," dubbed "The Giant Bathtub" by his workers, was 300 feet (east-west) by 155 feet and twice as high as any college bowl. It would be the first such structure in the history of bridge building.

The open sea of the Golden Gate is always rough. The derrick barge *Ajax* pitched so badly that her hands were often seasick. As for the deep-sea divers, in Strauss' words, "They worked at the constant risk of their lives." They could go below only during four 20-minute periods of slack water, between tides, each day. Even then, there might still be an ebb at one side of the Gate while the flood was already gathering on the opposite shore. To the eye, the 6-knot tides appeared much faster in the narrow strait. Pacific Bridge Company's J. E. "Jack" Graham later complained, "I hadn't figured on building a bridge in the *middle* of a river."

McClintic-Marshall Company's main divers were Bob Patching and John Bacon, assisted by George Nelson and Chris Hansen. For two years, they worked in inky blackness amid tugging currents and racing tides, setting charges to blast a great underwater quarry, scouring it clean with hoses carrying a 350-pound pressure jet stream of water, and fitting the 22 huge fender forms.

Patching observed:

> When we took off the wooden forms after the concrete had set, they went right to the bottom although they'd been as buoyant as cork three days before. That shows what the pressure is down there. An hour's about as long as you can stay at that depth. Longer than that is at your own risk. The deeper you go and the longer you stay, the greater the risk of the bends, or caisson disease, caused by over-rapid relief from the air pressure which has built up within the diving suit to counteract the water pressure without. The bends causes terrific pain, and sometimes temporary or even permanent paralysis in the part it hits you in. It gets me in the arms; it hits Bacon in the knees.
>
> A good tender is everything in diving. You have to trust him more than you would your mother. Your life depends on his keeping his head. My tender saved me from death once. The boom of a crane, above, broke and if he hadn't kept his head and yanked my air supply

Strauss' original plan for the Golden Gate Bridge, combining cantilever and suspension spans.
(California Historical Society)

A revised design, with a toll plaza styled after the *Arc de Triomphe*, Paris.

Finally, a more graceful design emerged. This interpretation was rendered by Maynard Dixon, a well-known landscape artist and muralist; this plan also featured an elaborate toll plaza.

hose out of the way, the boom would have fallen on it, and I'd have been permanently out of breath.

Strauss at first planned to tow a huge fender, like a caisson or floating drydock, to the pier site and sink it. He soon abandoned the idea because the top-heavy craft would be shipwrecked in the rough waters. He then decided to build his fender in units, great boxes 33 feet wide by 20 feet high, resting them on stepped benches blasted out of the bedrock. Within the curving concrete wall, he would sink a huge (but traditional) caisson and fill it with concrete to become a part of the pier itself.

First, the engineers had to run a 1,125 foot long (and 15 feet high) trestle from Fort Point to the pier site on the shoal. At its end, he would sink a guide tower, or guide frame, to which the first of his fender sections would be attached. He found Dr. Willis' "pudding stone" a nuisance; it was too hard to drill, so shoes for the trestle's bents, or legs, had to be blasted into the rock with dynamite bombs. The serpentine was so hard that it made necessary frequent replacement of the manganese steel clamshell buckets used in dredging. But tubular steel pilings were finally driven into the blasted rock, the bottoms of the bents inserted into them and secured with concrete.

The next headache for Strauss was a major one. On August 13, 1933, in a dense fog, the 2,000-ton freighter *Sidney M. Hauptman* plowed into the unprotected trestle about 400 feet offshore, carried away 300 feet of it and jammed the remaining six feet inshore. Divers removed the wreckage and the damaged bents and, to save time, replaced them with timber pilings. They were just in place when a storm roared in with waves 20 feet high and carried away the outshore bent—and the 50-ton steel guide tower. The first three box-like units of the fender were already on it, ready for one more on top and then the pouring of concrete. The tower, already weakened by the collision, could not support the load. It began to oscillate six feet backward and forward until the foundation pipes were displaced. On October 31, 1933, the whole structure toppled into deep water.

The destruction of his tower caused Strauss to change his strategy, several times. Finally, he decided to place a concrete seal within the fender up to the minus-64-feet mark. The caisson would sit on this flooring. He had the 8 x 30 foot steel guide frame salvaged, repaired and lengthened. But before it could be lowered in place again by the *Ajax* and a stiff-leg derrick on the trestle, another storm carried away 500 feet of trestle.

No wonder the engineer would later write:

> Without question, the most difficult engineering feat men have ever tackled was the south pier, rising 1,000 feet from shore on a rocky ledge 65 feet below the waves. I know of no place on the globe which has more violent conditions of water and weather than the Golden Gate. For eleven months it was an unequal battle of man against the sea.

The patient contractor rebuilt the trestle, using timbers in place of steel bents. The round piles offered less resistance to tidal currents than rolled steel "I" sections. The deck was raised to 20 feet to get it above the combers. Started on December 14, 1933, the new trestle was complete by March 8 and lasted out the job, four more years, without mishap.

Divers guided the legs of the tower into their sockets (pipe anchors) in the bedrock. To keep the sea from toppling it again, Strauss had a concrete block poured around its base. He also reinforced its fastenings to the trestle with heavy trusswork and ran wire rope cables as anchor lines, guying it in both directions. The lightweight steel frames ("boxes") offered little resistance to currents as they were slid down their rails. Divers fitted them with wooden panels weighted with concrete blocks to overcome buoyancy. Concrete pouring began on March 31, 1934. The forms tapered from 45 feet in width at the bottom to only 10 feet at the top. Section by section, the curving wall was built up, the units bound together with steel keys and reinforcing bars, but far below the surface at this point. Finally, a platform was built atop the ellipse at exactly the same height as the trestle. Whirley derricks moved in a circle, laying more concrete blocks. A second circuit brought 14 sections up to 15 feet above the water. The gap of eight sections at the minus-40-foot level was to allow entry of the $300,000, 10,000-ton caisson built at Moore Dry Dock Company's Oakland yard.

The bents and whirley platform removed, the caisson was neatly floated inside the fender, now as tall as a five-story building. Then, as preparations were made to close the open-

A 1931 drawing by Irving F. Morrow suggested the immense soar of the proposed south tower of the Golden Gate Bridge.

As the coastal steamer *Yale* passed the silent batteries of old Fort Point (built 1854-1861 and renamed Fort Winfield Scott in 1882), clamshell derrick operators both ashore and afloat off the precarious trestle began work on the footings of the Gate span.

ing forever, a tremendous storm broke, as though the sea had been setting a trap for Strauss. The turbulence inside the fender caused the caisson to surge, slamming into the curving wall like a battering ram. Fender frames were badly bent; the inner face of the concrete was deeply gouged. Even should the caisson be wrecked inside without doing further damage, it would delay—endanger—completion of the key to the entire bridge. Strauss had to make a hard decision, and fast. He later recalled the critical situation:

> We got the caisson in at 4 o'clock in the morning. It loomed up as big as a house. We edged it in very nicely and were all relieved. At 8 o'clock that night, I got a message to come out to the pier. Heavy swells had come in from the sea; the caisson was tossing around like a cork and acting like a ramrod and a catapult upon the fender. It looked as if it would batter it to pieces.
>
> It was necessary to make a quick decision and to find a substitute procedure which would be just as satisfactory. We finally hit upon a plan and, after discussion with the new contractor, arrived at a decision about midnight to adopt a new procedure and take the caisson out. This operation was just as ticklish a job as getting it in. It was successfully accomplished, however. The caisson was edged out about 9 o'clock in the evening, and later taken out to sea and sunk.

**E**ven this last action was not quite as simple as Strauss put it. The caisson broke loose from one towline. Coast Guard cutters and one tug stood guard over it while the other tug headed for home to repair its cable. There were fears that the caisson, with its 20 cases of dynamite, might become a menace to navigation. But the tug returned, got a line aboard and both took up the tow again. When 50 miles out at sea, the caisson was blown up and sunk.

There was one last narrow escape before Strauss finally overcame the jinx of the south pier. He recalled, "Early one morning in a heavy fog, after the fender had been pumped out, a great freighter loomed up only a few feet away. Had it hit the fender in that condition, all would have been lost. Fortunately, it just cleared, and we finished the pier without further alarms."

The chief engineer proved that his mind was as flexible as his future span. Very much in midstream, he again changed his plans. He converted his fender ring, completed on October 28, 1934, into an integral part of the pier. "We then poured a 40-foot mat of concrete, underwater, over the entire area inside the fender, pumped out the fender [November 27] and used it as a cofferdam and built in the dry." The pier was completed to its very bearing plates, ground "true" by a carborundum wheel, by January 8, 1935. On them Strauss would plant the two legs of the south tower, once Marin's tower was finished. (Strauss was using the same equipment for both jobs.)

Compared to the south pier, the Marin side was a piece of cake. A three-sided cofferdam was built against the cliff near the Lime Point lighthouse. The seaward corner was composed of cribbing, criss-crossed timbers forming cells. They were built upside down on barges and towed to Lime Point. Divers surveyed the bottom and the cribs were altered to fit its contours. They were then put in place, right side up this time. Broken rock was dumped into them and rock dikes extended to form a square. On a trestle built on the cribs, a crawler-type locomotive placed sheet-steel pilings all around the outside, like a board fence. It was not quite watertight, but steady pumping allowed the pier to be built "in the dry." Concrete was soon chuting down an "elephant trunk" on the pit's floor. Too big to pour at one time, the pier was built up in sections, each pour tied by shear keys and steel dowels to make a unified mass. The 64-foot high monolith weighed 45,000 tons.

On the San Francisco side, an old seawall was moved and pylons S-1 and S-2 begun on each side of old Fort Point. Between them was the Golden Gate Bridge's (hidden) satellite bridge, a graceful arch over the brick fortress.

The two cable anchorages were of the gravity type. Each was about 100 feet long and weighed 60,000 tons with long chains of 122 eyebars to grasp the 61 cable strands in pairs imbedded in the anchor block, which was sandwiched between a base block, bonded to bedrock, and a weight block.

The two 746-foot towers, one hundred ninety-one feet taller than the obelisk point of the Washington Monument, were built with two legs or columns, 90 feet apart, center to center, composed of 3½-foot square cells made of 7/8-inch thick plates and angles. These were of both carbon steel and the stronger silicon steel. As the towers tapered to their tops, the number of cells decreased, of course. Sections of the towers were hoisted by a vertical traveler, or creeper derrick, with two 90-foot booms with a lifting capacity of 85 tons apiece. The traveler, between the two shafts or legs, would place steel above its head then hoist itself up on top of it (via electric motors and tackle) taking 15 steps (or lifts) of 35 feet for the job. The traveler was supported in its new positions by four large pins or plungers, shot home like a deadbolt. The two massive towers (containing only a little less tonnage of steel than all four of the Bay Bridge's towers) were of tremendous strength because of their honeycomb design.

There were 90 designated "routes" through the mass of cells in the interior of each tower, via manholes and ladders. Workers often became lost, and two men had to spend an uncomfortable night in the eerie darkness. Strauss himself admitted, "Although I designed this weird labyrinth, I doubt if I could find my way out of it, even with the aid of the 26- page manual issued to direct the watchmen who inspect the towers over the 23 miles of ladders that connect the cells."

Once the one hundred-fifty-ton cast steel saddles were in place atop the towers, the cable spinning began in August 1935. The "laying of the gossamer web," as the *Literary Digest* lyrically put it, was quite similar to the Bay Bridge's technique. Skylines were strung for catwalks. For the long center span, the footwalk "ropes" (cables) were unreeled from a barge towed across the Gate, then hauled to their proper sag by hoisting ropes on the towers. Coast Guard boats, for the first and last time in history, closed the Golden Gate to all traffic during the ticklish operation.

The west footwalk was 15 feet wide, and on the east 18 feet because it carried an escalator-like endless moving rope. This "man-hauling" rope gave workers a boost as they climbed up the steep sag of the catwalks which, of course, had to imitate the eventual droop of the cables. Storm cables, with warning lights for ships, prevented the catwalks from whipping about in the Gate's stiff breezes. Fittingly for a bridge leading to the redwood coast, the flooring of the catwalks was of redwood planking rather than the cyclone fencing of the Bay span. Spaces were left to reduce wind resistance. Ten-foot panels of flooring were hoisted to the tower

tops in "nests" and then slid down the catwalk ropes to their proper position. Five cross-bridges stiffened the footwalks and gave easy access from one to the other. Overhead, two ropes supported the pipes on which the tramway rope sheaves were fixed for the spinning carriages.

Strauss, a zealot when it came to safety, not only installed handropes on the catwalks and had every fourth plank project a little as a cleat for sure footing, he also placed fire breaks at every tenth section of planking. A frame of light steel, with planks only wired on, could be quickly jettisoned to prevent the spread of a catwalk fire.

Wire from a storage yard at California City, near Tiburon, was brought to the site and fed to a double-wheeled carriage moved by reversible hauling ropes. Later, the carriage was fitted with four wheels, only three of which operated at a time (the lead wheel being "dead") so that three bights or loops of wire could be carried at a time. Each spinning wheel carriage shuttled halfway across the Gate, where the loops were shifted manually to the opposing carriage, then returned to its home anchorage with the exchanged bight of wire. At the anchorages, the loops were secured around strand shoes to form a bundle or strand. The tow sides of the loop were brought together at a "throat" just short of the shoe, which was pulled by a hydraulic jack until its hole lined up with those of flanking eyebars and a pin could be inserted.

The two 36-3/8-inch cables, the largest ever spun, were each 7,650 feet long and composed of 25,752 separate galvanized No. 6 wires, not as thick as a No. 2 pencil. The 61 strands in which they were gathered averaged 452 wires, but varied down to 256 in order for the flattened cable to lie better in its saddles and, when squeezed, to more easily assume a circular cross section. New techniques included the laying of strands in diagonal and vertical patterns, instead of horizontal, rows to minimize displacement of wires and slumping in the saddles. Cable formers, with vertical rods— separators—were used to lock the individual, loosely grouped, strands in proper position before compacting.

They set records for spinning, taking only three days to complete a four strand set-up. To avoid confusing the bights of wire on the sheaves, they had quick-drying red lacquer sprayed on one, green on another, and left the third un-

Behind the brick redoubts of Fort Point and inside the seawall on the San Francisco side of the Gate, earth movers began work on the deep-set anchorage.

An "elephant trunk" poised to pour concrete into the nest of reinforcing rods in the excavation, where the base block of the San Francisco anchorage would rise.

painted. The squeezing and wrapping machines were very similar to the yokes used on the Bay Bridge. The cables used similar splay collars and eyebars, too. The "tie downs" were in the concrete pylons. Pre-stressed suspender ropes were hung from cable saddles to support the deck's trusses, just as was done on the Bay Bridge.

The first suspenders were placed on August 31, 1936, and by September four traveler derricks were cantilevering truss panels out into space to be attached to them at every second panel point, that is, at 50-foot intervals. At 25-foot intervals, the derricks swung into place the crosswise floor beams to support the roadway. On top of them the travelers placed a few stringers, over which they traveled forward as they built. The other stringers, like curbs, paving, etc., would be added later in a "balancing act" to control dead load effects on cable pull and length, and on tower inclination. Diagonal members for wind bracing were then added to the stiffening trusses. As the truss panels progressed across the Gate, the concave sag of the deck under the travelers began to reverse itelf and finally assumed its graceful upward-arching camber.

The two suspension span derricks met at mid-Gate on November 20, 1936, and riveters joined the last two panels together at their top chords to end the "first pass" of steel erection. The lower chords were aligned by powerful hydraulic jacks until their rivet holes lined up perfectly. On the "second pass," the travelers laid the stringers, curbs, and sidewalk material. Seams and rivets and gusset plates were painted, as was any damaged surface. All suspended steel was in place by December 14, 1936.

Lack of cooperation among contractors cost three weeks of time during paving operations. Not till January 19, 1937, was the first concrete laid. The 7-inch slab, reinforced with transverse arc-welded steel trusses and by longitudinal rods, was poured in three 20-foot wide strips with transverse expansion points to keep it aloof from the stress in the trusses. Industrial locomotives hauled dump cars of wet concrete out on the deck to be distributed to the wooden forms by men wheeling buggies.

To allow the curve of the suspended deck to flatten during temperature rises, or under heavy loads, Strauss made use of expansion joints, rocker arms, and sliding pins in slots.

The expansion joints' fissures in the pavement were closed by sutures, grating-like bands of interlocking steel fingers. The largest pair, at the towers, each contain 500 of the metal digits.

Work wound toward its end as the approaches were finished and the sodium vapor lamps installed on the deck. The newspapers lauded Strauss and his "Queen of Bridges," realizing that the human element had been even more important than the tensile strength of cables or new alloys of steel. They saluted his winning combination of "brawn and iron." The chief engineer was the hero of the hour, for gambling on the greatest bridge in the world in the depths of the depression. But even the workmen felt possessive about the span; it was *their* bridge. Al "Frenchy" Gales, a Sausalito taxi and bus driver who switched to working on catwalks, cables, and footings for the Marin cofferdam, pier and anchorage, and both towers, told Stuart Nixon, "The whole deal was a marvelous experience for a country boy."

The Golden Gate Bridge was not only the most beautiful span in the world; it was the safest. There had been injuries from falls (The north tower elevator had even toppled over—with that tower's resident engineer inside!), but Strauss was proud that his care—and luck—had paid off. In 44 months of work, there had not been one fatality. He was no more superstitious than Purcell, but he was aware of the old high steel adage, "The bridge demands its life." In those years, that meant one death for every million dollars spent. But not till October 21, 1936, was his remarkable safety record tarnished when a derrick toppled and pinned a man, crushing him to death. Still, it was an isolated incident. There were other injuries, as when falsework collapsed on January 7, 1937, but the months rolled on without another death.

Strauss did not let up on his safety precautions. He was damned if he would accept such an "appalling sacrifice" of life, as tradition expected of his bridge. He had put a doctor and nurses on the construction wharf. When he suspected lead poisoning among the riveters in the tower cells, he had made them wear glasses and respirators and take physical exams, including blood counts, every two weeks. But, as Al Gales recalled, no one would wear the masks and goggles inside the towers. Strauss demanded clean hands, to prevent hand-to-mouth infection or poisoning. He changed from red lead to iron oxide paint on the splices of the San Francisco tower when he found that his worries about lead poisoning had been correct.

Joe Strauss insisted on safety belts. Francis A. Baptiste recalled 40 years later—"They were very safety conscious. I can still remember the large safety belt and the 3/8-inch, 10-foot long rope, always wondering if it was tied." Strauss was one of the first builders to insist on hard hats. They were primitive ones of leather, like the football helmets of the day, "antique" looking even in comparison with Purcell's. But they worked. (Ironically, however, the chief engineer would clamber over his bridge wearing his favorite cloth billed cap, while his assistant, Paine, wore a felt hat.)

The chief engineer put his "bridge monkeys" on special diets in hopes of counteracting tendencies in his men toward dizziness, giddiness, and vertigo. He equipped his tower workers with safety goggles, even tinting them like sunglasses because of possible snow blindness from the glare of bright days atop white fog banks. He fired men who drank on the job. He even served sauerkraut juice to workers after weekends, seeking the elusive hangover cure. Baptiste said the standard bridge joke was that it took 30 minutes to climb from the ground to the deck level of the tower before the outside elevator was installed—except on Monday mornings, when it took 40 minutes.

Strauss would not tolerate reckless showoffs. "To the annoyance of the daredevils, we fired any man who stunted on the job." He had no trouble replacing them, though he was supposedly confined to a labor pool drawn from the district's counties. (The poetic steelman, Harold McClain, later revealed that outsiders could and did pick up phony addresses and voting records in San Francisco.) Of his high steel men, Strauss mused, "One would think it next to impossible to find men to take the risks. But the very reverse is true. Bridgemen are a breed to themselves, strange migratory birds with an uncanny ability to sense the next big bridge job. There are always more men seeking work than there are jobs, [though] building a bridge is war with the forces of nature."

Strauss' major ally in his particular war with nature was his safety net, $82,000 worth of 7/8-inch manila hemp in a 6-inch mesh. (The cost was disputed; some said it took $130,000.) It was slung under the entire span and 10 feet out to each side beyond the width of the deck. Whatever its exact cost, the net quickly proved its worth. The first man to fall into it was carpenter George B. Murray, on October 16, 1936. He was knocked off the bridge by a car loaded with steel during a gale. On the 20th, Al Zampa slipped on a wet girder on the Marin side of the span. He broke three vertebrae and injured his pelvis because his weight sagged the net to the ground, only 25 feet below. In all, the net saved 19 men who formed the "We Fell Off The Bridge Club," which they shortly renamed "The Halfway To Hell Club." It was more exclusive, more elitist, than even the Bohemian Club.

As the work drew to a close, 25,000,000 man hours had been used, at a cost of but one life. It was incredible. As February 1937 arrived, and the bridge's opening was only three months away, there was little left to do on the span except such routine cleanup chores as removing the forms from the hardened concrete of the roadway.

On February 17, 1937, Evan C. "Slim" Lambert was bossing a 13-man crew on a traveling "stripper," a moving scaffold beneath the deck, suspended by steel rods or brackets from wheels running along girders of the deck. From it, workers stripped away the boards of the concrete forms and tossed them into the net where two of the men collected them. Slim Lambert, now a prominent Honolulu businessman, was a Bellingham, Washington, boy who had punched cows for a few years before turning to construction work on both the Bay Bridge and the Golden Gate span. He had just celebrated his 26th birthday, on February 11th. Lambert liked his job. He wrote Stu Nixon of the Redwood Empire Association in 1977, "Heights have never bothered me. Anyway, I was foreman and I really didn't have that much to do."

Unknown to Lambert, the unused-as-yet twin of his working platform had just been inspected—and condemned as unsafe!—at the south tower. At that very moment, State engineers were walking across the bridge to see his rig. It was just before 10 A.M. on a warm and sunny Wednesday of San Francisco's false spring. On the previous Monday, a State inspector, A.F. Mailloux, had questioned the safety of the 30 x 60-foot structure, and ¾-inch reinforcing bolts had been added to two of the four brackets from the wheels which advanced it. On Tuesday, Slim's crew had worked on it all day long without any trouble.

Work progressed on the Marin anchorage, situated on the headlands above Fort Baker's Horseshoe Cove, at the same time the San Francisco anchorage was being built.

Workmen waded in a morass of wet concrete to tie the stepped anchorage block to bedrock.

The concrete pylon holding the Marin anchorage was stepped like an ancient Greek theater, awaiting the concrete to seal the eyebars which would grasp the giant cable strands in place.

The men were chattering as usual as they worked, kidding each other about the soft, cushy job they had landed. Suddenly, the banter stopped. An eerie silence followed the sharp *crack* of a corner bracket. The entire apparatus tilted. Then a second corner tore loose, and as the entire weight of the staging and men bore on the remaining two, they gave way and the platform plummeted.

Immediately after the accident, Lambert told his story. He recalled the funny shudder and lurch of the staging. "I felt everything slipping. There was nothing to hang on to. So I hollered (to the rest of the fellows) and jumped into the net . . . I must have acted instinctively, because I don't remember thinking. I landed in the safety net. A moment later, I heard a sound like thunder as the 10-ton stripper ripped from its hangers. . . ."

Lambert hit the net just before the staging did. The hemp sagged slowly, ropes popped and the net fell away with another thunderous noise. But it did not drop quickly, but peeled away as if in slow motion. Recalled Lambert:

> Men were screaming and falling all around me. The whole net, about 1,200 feet of it, tore like tissue paper. I didn't realize until later that I was one of the few who missed being struck by the stripper. I was only conscious of being hurtled suddenly into the water. As I was falling, a piece of timber fell on my head. I was almost unconscious . . . I don't remember a thing except just before I hit the water with the net. Then I tried to jump. I think I succeeded because I wasn't fouled in the net.
>   I went down in the water, not very deep, I think, because I came right up again . . . The icy waters of the channel brought me to. I'm a strong swimmer and I tried to get clear of the rigging net and the wreckage . . . I saw some timber and grabbed on.

Nearby, Lambert saw three pairs of feet sticking out of the water. He did not know who the men were but he tried to get over to help them. He could not make it before the boots, pair by pair, disappeared. "They looked as though they were being sucked down. I closed my eyes for a minute and then looked up. About four feet from me was the face of a man I knew. It was like paste. I tried to swim toward it. Then a horrible expression crossed the face. I knew the man was dead. He sank before my eyes." (Originally, Lambert thought that the man was a friend of his.) "There, tangled in the net was Noel Flowers. I yelled, asking could he cut himself out. He just looked at me. God! What a horrified look! Then he went down." (It must have been a case of mistaken identity. The name was not among those listed as missing.)

Looking around, Lambert spied two more feet. "I swam toward them, as fast as I could; I pulled the body up and got him in my arms. It was Fred Dummatzen. For what seemed like hours, I supported his body. It was a deadweight. I was afraid that he might be dead, but I couldn't take the chance. I thought that if we were rescued quickly, they might be able to save him. . . ."

Lambert drifted out through the Golden Gate, his body like a block of ice. Blood ran from his head wound into his eyes. All around him, he felt, were the bodies of his buddies, and he could do nothing for them. "I think I was in the water about 45 minutes. It was 'way too long." (It seemed to be an eternity to Slim, but it may have been only 20 minutes.) "A Coast Guard vessel came out, trying to find survivors, but it didn't come far enough to sea to get me. We were out by the [Point Bonita] lighthouse." The cutter closed to within 100 yards of Lambert in one pass, but failed to see him because he was hidden by the debris being carried seaward by the ebb tide.

Finally, a crab fishing boat, returning to port, rescued Lambert just as he was about to give up. Slim still had an arm around Dummatzen, though his pal was dead. Of the fisherman (whose name he forgot), Lambert said, "He had a hell of a time getting us into that boat, I can tell you."

Amazingly, Lambert was not the only survivor of the 200-foot fall. A 51-year old carpenter, Oscar Osberg, was fished out of the sea, too. He was alive though badly injured. His right hip and thigh were fractured, his ribs broken, by falling timber. The bodies of the other ten men vanished in the sea.

When Lambert came to at Mary's Help Hospital, suffering from immersion (exposure) and shock, his first words were, "How many of those men got killed?" He later remarked "It's like a terrible dream. It couldn't have happened. Yet I know it did. Ten of my friends were dead. I saw them die all around me and couldn't do anything about it . . . I didn't remember much more. I don't know any of the details of what hap-

The huge eyebars would hold the bridge cables fast. In the right background, across Horseshoe Bay, is Point Cavallo.

The enormous mass of concrete which welded the Golden Gate Bridge to the Marin headlands at pylon N-2 filled the nest of eyebars to their very heads.

The anchorages on both sides of the Gate were completed fairly quickly. Then work on the piers—the tower foundations—began. The north (Marin) pier was finished eight months ahead of its San Francisco counterpart.

pened. I only know I keep hearing the cries of my friends. They were frightened. It's hell to die like that."

In 1977, Lambert mused, "I have always felt fortunate in having survived such a terrible and tragic accident. This adds to my feeling of extremely good luck." But the foreman was as gutsy as he was lucky. His concern in the water had been to save others as well as himself. He did not dwell on his own personal shock and pain. And he was back on the job, high above the treacherous Gate, in less than two months, though he had suffered a broken shoulder, ribs and neck, the latter injury unknown to either doctor or patient. Slim explained, "They found the broken shoulder and ribs at the time, but they only X-rayed me from the neck down and I didn't know, until years later, that I had broken several vertebrae in my neck."

The rending of metal, the cries of the doomed men, the explosive popping of the hemp chilled the blood of every worker within sight or hearing. And, according to the *Chronicle*, the "crackling, snapping, whipping sound [was] heard far into Marin County."

Some men claimed that the whole bridge shook as the netting was torn away. Steel worker Tex Leaster, high up on the top of the south tower at the time, said, "I felt the tower tremble as though there was an earthquake. I could see the net falling, accompanied by a sort of subdued chatter. I could hear faint, baby-like cries. When the net hit the water, the men seemed like little blots of ink on the surface. The net looked like a rapidly sinking raft, with the tiny men entangled, fighting to get free. Then some of the inky blots disappeared. Some drifted out the Gate and out of sight."

One man grabbed the dropping web of hemp and swayed, suspended in the air, until he lost his hold and dropped. In the water, he evaded the net which, designed to save lives, was now a trap in the sea. He swam for some minutes, but then his comrades on the deck above saw his face disappear from sight under the waves.

Slowed by the net and buoyed by the ocean breeze, the men did not seem to fall as much as to float downward. But some were screaming in fright. When the net hit the water it churned the Gate—yellow from river silt—into a lathered foam, hiding the men entangled in the hemp. One or two broke free and swam for shore, but did not make it in the cold water.

Roy Trimble, a painter, was much closer to the accident than Tex. He remembered, "The cries of the men seemed to split the air. The weight of the crane and falling net shook the bridge like an earthquake. There was a 10 to 15 mile west wind and the net waved like a huge flag as it carried the men down, away from us. I shall never forget the contorted white faces fading away."

John Marders, stripping scaffolding but atop the deck, said, "When the scaffold hit the net it sounded like a clap of thunder, followed by the rat-tat-tat of a machine gun, only louder. When the net started to rip away, I got so excited I kneeled on the bridge floor—and reached down. I guess I was out of my head, making a futile gesture to save those poor fellows. I almost fell off; someone pulled me back. The snapping of the rope that held the net mingled with the screams of the men. The noise was indescribable."

The scaffold, suspended by a crane and hand-winched along on 6-inch wheels, was a wooden frame covered with planks. But the frame was attached to steel beams. Frank Dowling, a timekeeper, reported "There were two men clinging to the great steel beams that were diving into the water. It looked as though they were holding to the steel for safety! Two other men were clinging to the net and struggling, as though trying to climb up the falling ropes."

Some workers ran along the deck, hunting for life preservers to hurl over the side into the tide tip. There were fire extinguishers, but no life rings. Others shouted to fishing boats and gestured to point out men swimming for their lives. Most of the workmen stood helpless. Peter Anderson had to watch his own brother fall to his death. Whistles blared an alarm and all work, for the first and only time, came to a stop. The inspectors, A.F. Mailloux and others, hurrying to examine Lambert's staging, heard the crash but did not see the fall. Incredibly, one man captured the net's collapse with his camera. He was Joe Dearing, later a sports photographer for the San Jose *Mercury*.

Everyone crowded the edge of the deck to watch the Coast Guard boats and fishing boats circling below in a search for survivors. Coast Guardsmen arrived in small boats from the Golden Gate and Point Bonita stations in just 12 minutes and were soon followed by the 75-foot cutter *Tahoe*. Edward Riley, foreman painter, recalled, "After the net hit the water, I saw one swimmer reaching out to help another. The second man, terrified, struggled with the first, and both went under."

The tragedy was unforgettable. The shock of it remained in the minds of its viewers long after other bridge memories were gone. Not surprisingly, the number of "eyewitnesses" grew with the passing of the years. Walter "Peanuts" Coble observed in 1977, "I doubt that everybody who says he was there, February 17, 1937, when the scaffold fell, could have been. Otherwise, the Bridge itself would have fallen under their weight."

It was not known which men had fallen until all 200 workers on the span were dismissed for the day. When, after three hours, only ten time cards remained unpulled, they were collected and checked—Dummatzen, Anderson, another Anderson, Bass, Desper, Halliman, Hillen, Lindros, Norman, and Russell. One of the victims, the *Chronicle* was told, had often reassured his worried wife, "It is one of the safest places to work in the country."

All but two of Lambert's crew rode the net down. Wayne A. de Janvier was admitted to the "Halfway to Hell Club" along with Tom Casey, though neither qualified technically in that they did not go into the netting. When he felt the jar and quiver of the breaking platform, de Janvier did not jump or fall like the others. His reflexes told him to grab something so he clung to a girder for dear life. In 1977 he quipped, "My fingerprints are still imprinted in the steel beam where I hung on till rescued."

Laconic C.H. Thompson, a painter, admitted to a *Chronicle* reporter, "I guess I helped save a man from death. He was Tom Casey, the little red-headed carpenter." The moment he heard the rending of metal and felt the platform canting, Casey instinctively grabbed a caster overhead and held on for dear life, like a limpet. He dangled in space for seven minutes. And, as he later said, "It seemed a hell of a lot longer than that to me!" It took his pals about three minutes to throw him a line. He spurned it. Later, he explained his action to his mother. "The first one that came down was thin and greasy. There were no knots in it. I figured I might as well let go as try to shinny up that, so I hollered up, 'Send down another!' So they did." Quickly, William Foster, a State inspector, with some riggers made a loop in a line and dropped it. The Irishman inserted a leg in it and was pulled to

safety. With Gaelic stoicism (or perhaps Irish frugality), he still had his pipe clenched tightly in his teeth! Casey's first heroic words? "Gimme a match."

Casey was told to skip work on the 18th to get over his fright. The *Chronicle* ran a story on him (and his pipe), complete with photo. He used his day off to write his mother in Ireland: "The luck of the Caseys, Mother, it was." (Like many Irish, he tended to exaggerate a bit, so he threw in an extra 80 feet of height as he went on . . .) "And me, 300 feet up, hanging on like a monkey . . . A Casey never lets go, huh, Mother? But your son could not have hung there much longer . . . And I never let go my pipe . . . I had her in my mouth all the time."

Two hours after the accident, as Pan Am's China Clipper flew overhead, the wooden framework of the traveling stripper rose to the surface, most of its planking gone in the rip tide. To reporters, it looked like two giant picture frames. Ashore, the grief-numbed families of the victims had come in taxis and family cars. They gathered around the Pacific Bridge Company's timekeeper's shack with their forlorn hopes.

On the 19th, the net was found with grappling hooks. It took most of the night for the powerful derrick barge *Haviside No. 4* to raise it. One body was recovered.

Four separate inquiries into the accident were held. De Janvier was a surprise witness at the San Francisco coroner's inquest. Referring to the side brackets which Mailloux had wanted strengthened with extra safety bolts, he said that he had told Shorty Bass, a rigger, "It doesn't look a damn bit safe to me." The Bridge District had an inquiry, as did Pacific Bridge Company, and the State Industrial Accident Commission's often acrimonious hearing was broadcast on radio. Metal failure was the cause of the tragedy.

Although some unconcerned painters swung from Jacob's ladders, painting suspender ropes, most work on the unprotected south half of the span was halted until new netting could be slung, starting on March 3 and completed April 15. Strauss, like a symphony conductor, had to coordinate the delayed work of at least three contractors in the clutter of the deck. Many of the workers were drifting away to new jobs. (It was probably their restless nature more than the accident.)

The rule that workers should be from the district's count-

As men worked on the trestle to create a secure foundation for the south tower, the north tower progressed across the Gate.

Even the steep cliffs of Marin County were dwarfed as the north tower rose skyward.

ies (often honored in the breach, in any case) was now abandoned. Wrote Strauss: "Since all operations on the bridge at this time began to feel the scarcity of skilled workmen of all crafts, the rule for employment of residents only was waived, when necessary, to minimize delay."

Equipment and debris was removed from the deck as a final cleanup began. Splatters of grout and cement were removed with putty knives and steel wool, and surfaces touched up with an undercoat before the top coat of "International Orange" paint was applied.

The men who had lost their lives were not forgotten. Besides memorial services, a plaque was placed on the San Francisco abutment. It reads in part: "For forty-four months out of a total construction period of fifty-two, tragedy passed the Golden Gate Bridge by. Then death struck twice, claiming the lives of eleven builders of the bridge . . . To these dead, who made the supreme sacrifice, the living pay tribute for their contribution to California and that which she has achieved. They gave their all. None could give more."

Purcell had driven home the last of the Bay Bridge's rivets. But Strauss let Shirley Brown, daughter of Golden Gate Bridge Fiesta (May 27-June 2) chairman Arthur M. Brown, Jr., do the honors. The symbolic last rivet, a gold one donated by Charles Segerstrom of Sonora, was then withdrawn and a real rivet driven home by Edward Stanley, who had driven the bridge's very first rivet. The job was done.

The Fiesta which, in a way, saluted the completion of both bridges and progress on Treasure Island's World's Fair (1939-1940), the Golden Gate International Exposition, began at 6 A.M. on May 27, Pedestrians Day. Sprinter Donald Bryant led 18,000 people in a race to be first across. By nightfall, 200,000 people had crossed, including this writer. Everyone came, it seemed, except suicides and Dr. Bailey "Pudding stone" Willis. The 80-year-old professor was too busy, hunting for gold in the Philippines.

Surprisingly, only 32,000 cars crossed on the first auto day, May 28, in spite of the oratory, pageantry and parades—including airplanes overhead and warships beneath the span. At night, Strauss' "eternal rainbow" of steel was lit by the Pacific Fleet's searchlights in a sort of early *son et lumiere* (without the sound.)

The poetic Strauss and his company received $1,080,000 in pay for building the "impossible" bridge, but not a penny for his poetry about the span:

> Launched midst a thousand hopes and fears,
> Damned by a thousand hostile sneers,
> Yet ne'er its course was stayed.
> But ask of those who met the foe
> Who stood alone when faith was low,
> Ask them the price they paid. . . .

**S**trauss, in a sense, paid the price like the eleven victims of the bridge. His health broke in 1937, probably due to his long exertions. He participated in the completion of the bridge but he died of a heart ailment just a year later.

The district had Frederick W. Schweigardt execute a bronze statue of Strauss for the toll plaza in 1941. Traffic was stopped for several minutes of silent tribute to the man of genius who was said to have erected 400 bridges in his life, one of which made him immortal. The "curve of soaring steel, graceful and confident over infinity," of poet Henry May's lines, has been called "The Wonder Bridge of the World."

In his final report, Strauss acknowledged the old truism that every bridge contributes to the science of structural design and building techniques. But he stressed that the Golden Gate Bridge's major contributions were in architectonics. Its beauty transcends its magnitude. Choosing the simplicity of the suspension design, he added well-proportioned, portal-braced towers with a set-back design which gave an exquisite taper and simplicity of line to them while imparting what he called "reserve power and serene dignity."

Irving Morrow, the architect for the Golden Gate Bridge, brought the tops of his pylons up and made symmetrical monuments of them, using the vertical set-back motif in their concrete, just as in the steel towers. Some critics dislike the "modernistic" Art Deco-like touches, which extend to light standards and hand railings. He used structural steel shapes without ornamental elaboration. He was innovative, reversing the usual architectural practice of enlarging the scale of de-

(Stephen W. Fotter)

tails (on his towers) as the distance from the ground increases.

Morrow sensed that pedestrians would claim the bridge as their own, so he flanked the roadway with wide sidewalks whose attractive open handrailings and offset bays—lofty vantage points— encouraged gazing and photography. Where the towers blocked the sidewalks, he wrapped balconies around them to create unbroken promenade decks.

Like John Ruskin, Joseph Strauss had an appreciation of "the poetry of architecture." For this reason, he used X-braces only below the deck, where they are seen by few persons.

For the thousands who drive and stroll the bridge, he braced his twin tower columns with simple horizontal struts which he sheathed in steel fascia plates. The parallel lines, a simple fluting of intersecting plane facets, catch and hold the eye and relate it to the verticality of the flanking columns. Such touches as these make Strauss' bridge his masterpiece, as beautiful as it is strong and dignified. His other 399 bridges were just practice for the Golden Gate Bridge, in which he achieved a sculptural elegance and a spiritual eloquence unrivaled in the world.

The creeper derrick squatted between the legs of the north tower during its slow ascent. So immense was the north tower that workmen on it, like the man stooping over on the lowermost staging, almost vanished from the camera's eye.

The north tower, its creeper derrick still at work, surpassed the old guardian of the Gate, Lime Point Lighthouse.

The south pier was still under water at the end of the storm- and ship-battered trestle off Fort Point as the Marin tower and the pylon, which would carry cables to the San Francisco anchorage, neared completion.

Victory! The critical south tower pier rose at last above the waters of the Gate, insuring the successful completion of the bridge.

At last, the south tower
caught up with its twin,
and both awaited the
saddles into which the
giant cables would nestle.

Like the Pillars of Hercules, the towers of the bridge stood guard over the Gate in the summer of 1935.

In preparation for cable spinning, the twin towers, 4,200 feet apart, were linked by ropes to support catwalks.

An aerial view of the Golden Gate as cable spinning began showed the components of the span—the Marin anchorage pylons (extreme left), the north tower, the main span and south tower, the temporary trestle, and San Francisco's concrete pylons and anchorage.

The spinning carriages were made ready for their many trips halfway across the Gate from their home anchorage, as eyebars (lower right) stood ready.

The catwalks were secured against the gale winds of the Golden Gate with strong cables.

Redwood-slatted catwalks stretched across the Golden Gate as a ship called on the port in December, 1935.

The cleated redwood cat-walks were soon paral-leled by banded strands of the growing cables.

Engineers often met to confer on the catwalks of the Golden Gate Bridge.

A silvery sunset transformed the cable spinning operation into photographic poetry; the spinning wheel moved at 640 feet a minute as it spun its web.

From a crossbridge, cameraman Moulin aimed his lens at the south tower on a December day in 1935.

The photographer then swung about on the crossbridge and shot Marin's north tower.

Moulin's shutter "stopped" the west spinning wheel, seemingly supported by guy wires and sky hooks, as it left the south tower.

Close up, the entire carriage of the spinning wheels was taller than a man.

The shuttling wheel was paying out wire at the top of the north tower when Moulin next captured it in action. The strands were held in temporary positions before being dropped into the saddles to form a finished, round cable.

Strauss could not count on engineering alone to bridge the Gate. He had a tremendous asset in the caliber of his workers, all of whom seemed to adopt the bridge as their own. The press liked to attribute the completed bridge to "brawn and iron."

*Center page.* The sagging cable strands and catwalks appeared to almost touch the water, but there was over 200 feet of clearance for the *President Hoover* of Dollar Lines.

Cable strands had to be banded against loosening ("unraveling") to keep them compact and circular.

117

As two "leather hats" guided cable strands into position, the camera moved in for a closeup at the very top of Marin's north tower.

The north tower during cable spinning. At the top the saddles would accept and grip the completed cables.

Above the workers with the spinning carriage and its bights of wire, the twin catwalks climbed to the north tower at a dizzying angle.

The anchorages on either side of the Gate, a jumble of cables, eyebars and sheaves, stood ready to receive the cable strands.

The loosely grouped strands were almost ready for compacting into the strongest cable on earth.

Past pylon S-1 sagged the strands which would soon be bound into unified cables.

121

The major difference between the banding of the Bay Bridge and the Golden Gate Bridge cables was that the crews of the former's squeezing machine wore metal "hard hats," while the latter's affected antique leather helmets.

Even when half finished, in 1936, the Golden Gate Bridge was so graceful that most of Strauss' critics were silenced.

The first Marin County commuter? No, just a lonely engineer or inspector, making his rounds.

As the stiffening trusses advanced from the towers, Joseph Strauss' famous safety net made its first appearance.

The Alaska packer, *Star of Zealand,* bade farewell to San Francisco on September 19, 1936. Under tow until she reached the open sea, she then made sail for Japan—and a wrecker's yard. Workmen on the north tower were among the last to see the proud old ship.

Workmen installing the stiffening trusses of the bridge clung to smooth steel girders.

Every moment of every work day saw little dramas unfold in the placing of each truss, but the men's movements were more ballet than conventional theater.

127

Piece by piece, like a giant jigsaw puzzle, the stiffening trusses were hung for the deck.

Without benefit of safety belt (or parachute), workers placed their faith in their balance—and Strauss' net below, which was cantilevered out ahead of the work.

Even on a hazy day, the "high iron" men fitting the deck trusses had a splendid view of the "rival" bridge to the east.

Long hours were common on the Golden Gate Bridge, though the official work day was just eight hours (compared to the Bay Bridge's six-hour day.)

Great crossbeams were maneuvered into place by traveler derricks—and muscle power—to take the roadway of the deck.

The safety net beneath the crossbeams cost no less than $80,000—and probably more (to cover the entire bridge, finally)—but it was worth it. Nineteen lives were saved by it. These lucky workers promptly organized the "Halfway to Hell Club."

As with the Bay Bridge, trusses were placed in a strict order, leaving gaps in order to avoid concentration of weight and stress.

"Brawn and iron" were typified by the tightening operation on the bolts under the cable bands, from which the suspender ropes hung.

Strauss' net, looking like gauze beneath the trusses, was actually mesh woven of 3/8-inch manila rope.

The frame of falsework hid the beauty of the arch over Fort Point, the hidden "bridge within a bridge."

136

To the deck floor were added diagonal braces and stringers which would support the concrete roadway to come.

Only a few stringers had been laid on the deck to carry the traveler derrick when this shot was made from high up on the south tower.

The concrete of the pavement was poured in patches, again to equalize weight and stress over the entire bridge. A temporary railroad track brought material and fresh cement to the pour points.

The Marin landing was a great pile of litter as paving of the roadway progressed.

The roadway was poured and workmen were stripping away the wooden forms beneath the pavement when tragedy struck. The scaffold on which they were working beneath the bridge collapsed and tore through the safety net, falling into the Bay.

When the stripping scaffolding, carrying ten men with it, tore away in February, 1937, it ripped away almost the entire safety net—1,200 feet of it.

Fishing boats came to the rescue of the men who fell into the bay, but were able to pull only two out of the water alive.

Captain William Beechey of the Royal Navy, who named The Needles (the pointed rocks near Lime Point), would not have recognized the Golden Gate in 1937.

At long last, the bridge was "within an inch" of being finished; only forms on the concrete pylon and the ubiquitous derricks of the tower tops needed to be removed.

A close-up of the string-
ers and the diagonal
bracing of the deck.

At this stage, it appeared to casual observers that the Golden Gate Bridge would *never* be finished.

*Opposite page.* At dusk, the traveler derricks met at mid-span as if to symbolize the completion of the span.

The bridge as seen from behind the Coast Guard dock, just north of Crissy Field, was almost ready for its debut.

A vista of the new bridge taken from an old gun emplacement above Fort Point.

The debris was cleared away; the bridge was ready—and not a soul on its deck.

The bridge made a new and impressive backdrop to panoramas of the City, especially from Pacific Heights.

The bridge before it opened was the center of this mood study—eerie, empty, fogbound.

Fishermen at Fort Point soon learned they had one of the best views of the beautiful Golden Gate Bridge.

The dream of every
motorist—the entire
Golden Gate Bridge to
oneself—came true for
this engineer or inspector
or official, as the bridge
was cleaned up for its
May, 1937 opening

The sodium vapor lamps of the bridge were bright, but not bright enough for the Golden Gate Bridge Fiesta nights, celebrating completion of the span. Klieg lights, spotlights, searchlights—all joined in the fun of celebrating the completion of Strauss' "eternal rainbow" of a bridge.

Another view of the deserted bridge, a day or so before the official opening—May 27, 1937, for pedestrians only, and May 28 for autos.

A graceful curve of lights beckoned San Franciscans to "marvelous Marin."

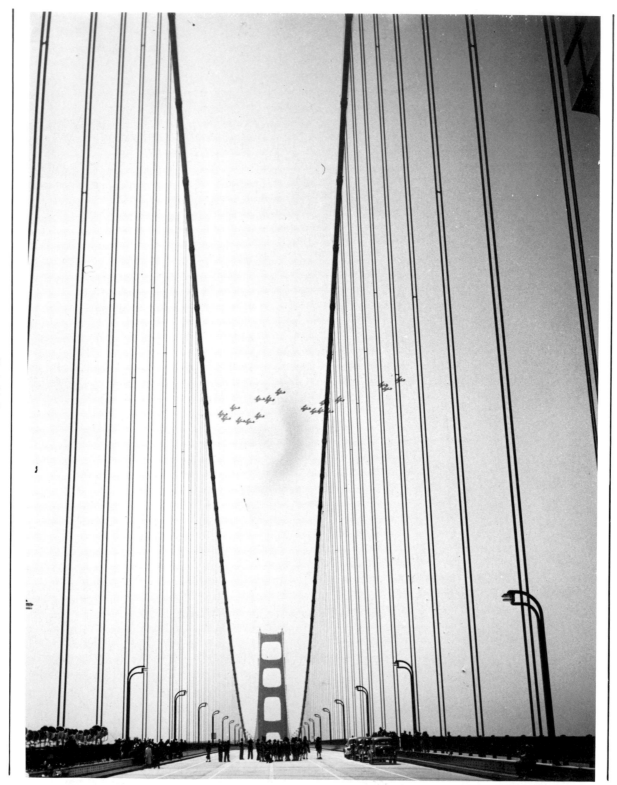

As the bridge was about to be opened for Pedestrians Day, combat biplanes flew overhead in salute.

May 27, 1937, was Pedestrians Day, reserved for foot traffic. Some 200,000 people had crossed the Golden Gate Bridge by day's end.

Fog did not dampen the spirits of the thousands of people who filled the bridge, hour after hour, from railing to railing on Pedestrians Day. The next day, May 28, 1937, was the first day for autos to cross the bridge. Fewer came than were expected. (Perhaps many drivers had crossed as pedestrians the day before.)

163

For more than 40 years, airline travelers have viewed the beautiful contours of the Golden Gate and San Francisco Bay bridges, yet they are always fresh and new, never failing to thrill a resident or extend a welcome to a visitor.

By 1938, San Francisco Bay had assumed its (presumably) permanent appearance, with the two new bridges and Treasure Island gleaming in the sun.